astronomy

HOW MAN LEARNED ABOUT THE UNIVERSE

Lou Williams Page

ADDISON-WESLEY PUBLISHING COMPANY

Reading, Massachusetts • Menlo Park, California • London • Don Mills, Ontario

Second printing, August 1971

Copyright © 1969 by ADDISON-WESLEY PUBLISHING
COMPANY, INC., Philippines Copyright 1969 by
ADDISON-WESLEY PUBLISHING CO., INC.

ISBN 0-201-05650-X
BCDEFGHIJ-AL-7876543

To Thornton Page,
in gratitude for his wisdom
and patience

Preface

The same night sky looks down on man preparing to visit Mars as looked down on the shepherds of ancient Greece. They saw Mars only as another bright jewel among the many that spangled the heavens. Centuries later we know that this small light in the night sky is big enough and solid enough to land on, and that its lack of water and oxygen and its daily extremes of temperature will make it inhospitable to man. Mars' path through the trackless reaches of space is so accurately charted that its location at every second is known. Yet the closest we on earth have come to Mars is about 35 million miles. How did man learn so much about it, even before Ranger VIII took a closer look?

Astronomers tell us that the earth is moving around the sun, and yet the planet certainly appears to be motionless. They tell us that the earth is one of nine planets moving around the sun, but no one has yet looked down on their orbits, drawn so confidently in your textbooks. They tell us that the sun, almost 100 million miles away, is the nearest star, one of the billions gathered together in a turning island in space called the Milky Way galaxy. They tell us that there are at least a billion other galaxies, each crowded with stars, and that many of these stars may possess solar systems of their own—and that some of these solar systems may include a planet like our earth, capable of supporting life. They tell us that some stars are old and some are young, and that even now we are witnessing the birth of stars. They say that some stars are hotter than others, that all of them can shine by means of nuclear reactions going on inside, and that the chemical elements we find on earth are the same as those in stars throughout the universe.

Go out and look at the silent and remote stars; see exactly what ancient man saw. How has this tremendously detailed story been read from these points of light? Why or how did men stop thinking of them as only points of light?

Isolated on the planet which is our home, we can only look out at the universe and speculate about it, like passengers marooned on a ship at sea, staring at the distant shore and trying to make out its features, its nature, and its distance. In the past decade we have launched satellites and space probes into nearby space, like tiny lifeboats leaving an ocean liner for a somewhat closer look.

For thousands of years men have gathered observations of the universe, and some have pieced these bits of information into a meaningful picture. This picture has changed, greatly or in detail, with each new observation, with each new mechanical aid, with each scientist who was willing to look at old evidence with new eyes. While the universe has remained much the same through the ages, our ideas about it have changed over and over again. Each accepted theory was the best picture men could draw on the basis of observations made up to that time. Year by year, new parts of the picture are being "roughed in," and other portions are being painted with new and sharper detail. Unsolved problems will always remain. A great deal of the canvas is yet untouched, waiting to be filled in by astronomers and astronauts of the future.

Lou Williams Page

Contents

chapter 1 | **The Silent Stars**

If it's clear tonight, go out and look up at the sky. If there is no moon, all you will see are jewel-like points of light, some of them lying in a misty band across the sky. You can count two thousand stars if you have good eyesight and live in a dry climate, far from city lights. You will see many, many less if it is a moonlit evening and you are in town.

Above the earth will be the same sky that looked down on Caesar's legions, on Greeks strolling across the newly built Acropolis, on shepherds tending their flocks while David wrote the Psalms. But these people's thoughts about the starry sky were very different from yours. To them the stars were merely what they appear to be: sparkling decorations that make the dark, bowl-shaped sky beautiful and somewhat awesome.

While you may appreciate the beauty of this heaven full of stars as much as they did, your thoughts about it range much farther. You have heard that the stars are huge and hot, that their numbers and their distances are so great that they defy the imagination. Questions come to your mind: How do we know that most of them are glowing spheres of gas? How do we know that some are old and some are young? How did people become convinced that some of these tiny lights are much nearer than others and cool and solid enough to land on? How can we pick these out from their many neighbors in the sky? How can we make intelligent guesses as to whether there are living things on any of them? Can the stars tell us how the universe began? Most of these questions were not thought of, much less answered, by the Romans, the Greeks, or the Egyptians before them.

The stars have remained silent and man has remained isolated from them on the far-away earth. Yet, two thousand years and a million stargazers later, there are answers to these questions and to many more. Astronomers have used their eyes, aided by instruments that they have devised and made; and they have reasoned about what they saw. Each hard-won answer made our picture of the universe more complete and each new answer raised new questions. There is an unbroken line of astronomers from the first man who wondered what the stars are and tried to find out, to the astronauts' successful journeys into nearby space and the current, hotly disputed theories of the universe's origin. Let us join that line and begin to look and reason, starting where the first astronomers began.

If you face north, you will see the stars that are shown on the inside cover of this book extending almost from the horizon to the "top of the sky." This point in the sky, directly overhead, is called the *zenith*. When you first look at these stars, in the drawing or in the sky, they seem to be a random scattering. But if you look at them often enough, or hard enough, they will seem to form patterns. To the young heroine of Ludwig Bemelman's *Madeleine,* in the hospital recovering from appendicitis, "... A crack in the ceiling had the habit of sometimes looking like a rabbit." And so, to the ancients, who spent far more time under the open sky than we do, groups of stars seemed to form the figures of people and animals. They called these groups of stars *constellations.*

Let us look first at the Big Dipper, a star group that you probably know already. If you have any trouble finding it in the sky, have someone point it out to you. Turn the book so that the drawing of the Big Dipper is at the same angle as the Dipper in the sky. Then the other constellations shown will all fall into place between horizon and zenith. See if you can pick out all five of them.

The Big Dipper is a convenient yardstick for measurements in the sky. The heavens seem like half of a sphere viewed from the inside. The horizon is a circle and therefore measures 360 degrees (360°) all the way around; while the half

circle from one horizon, up the sky to the zenith and down to the opposite horizon, measures 180°. We remember that each degree is divided into 60 minutes (60′) and each minute is further subdivided into 60 minutes of a minute (60″). This second minute is more simply called a *second.* The space between the two stars forming the side of the Dipper opposite of its handle measures just about 5°. A line joining these stars, if continued for almost 35° (almost seven times the distance between these "yardstick stars"), reaches a star called Polaris, the end star in the handle of the Little Dipper (Fig. 1-1). Cassiopeia can be found by drawing a line from the end star in the Big Dipper's handle, through Polaris and, about 20° farther, to the end star in Cassiopeia. So, when you have found the Big Dipper, the other star groups in this part of the sky should be easy to locate.

The Big Dipper is not really a whole constellation. It is part of the constellation of Ursa Major (the Big Bear), which includes other nearby stars. The Little

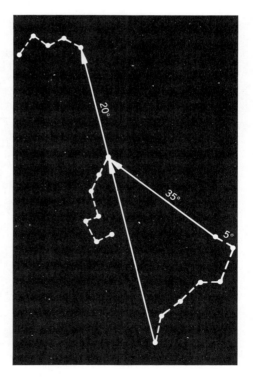

Fig. 1-1 Finding the Little Dipper and Cassiopeia by means of the Big Dipper's "yardstick stars."

Dipper is also part of a larger group of stars, the constellation Ursa Minor (the Little Bear). Both bears are rather difficult to discern, and for our purposes the clear and plain outlines of the two saucepans are more useful. Draco the Dragon, Cepheus, and Cassiopeia are more complete constellations, however, and show how the ancients used the sky as a sort of historical and religious picture book. The stars of Cassiopeia form a clear although shallow "W," an alphabetical symbol unknown in ancient Greece. So, by including a star just above the point of the "W," they made it into a chair (Inside Cover) — in fact, into the throne of Queen Cassiopeia of Ethiopea, banished forever to the sky by the gods because she boasted too much of her beauty. Her husband, King Cepheus, is there too, looking to us like

Fig. 1-2

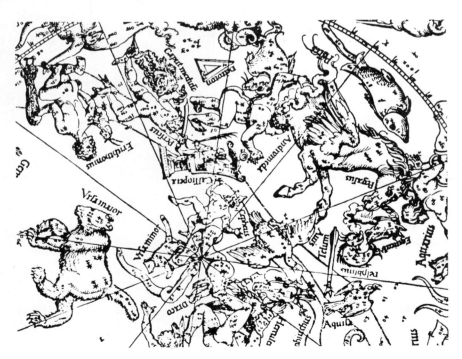

Fig. 1-3 Albrecht Dürer's 1515 representation of Ursa Major, Ursa Minor, Draco, Cassiopeia, Cepheus, Perseus, Andromeda, and Pegasus.

a country church complete with steeple. But some fanciful stargazers saw him as he looks in Figure 1-2. In fact, as constellation drawings go, this is quite a simple clothing of the bare stars! Especially during the Middle Ages, but also until the 1850's, many well-known artists made imaginative and elaborate drawings of the constellations. Figure 1-3 is part of a sky map made in 1515 by the German artist, Albrecht Dürer (perhaps best known to us today for his woodcut of hands folded in prayer). It shows the five star groups you have just seen in the sky, together with others near them.

Equally elaborate as these drawings were some of the ancient stories about the constellations and the adventures of the people they represent. Cepheus, Cassiopeia, their daughter Andromeda, son-in-law Perseus, and even his horse Pegasus (Fig. 1-3) are all immortalized as constellations and are the subjects of an involved 4000-year-old adventure story. Becoming familiar with all the constellations and their legends would require a program of full-time study.

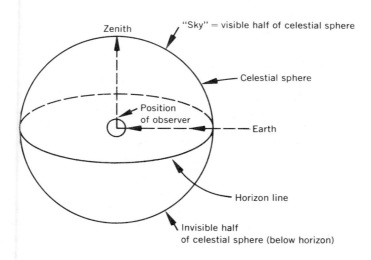

Fig. 1-4 The celestial sphere, horizon, and zenith.

From horizon to zenith, the bowl or half-sphere of the sky is studded with constellations. Even the earliest astronomers saw no reason to suppose that there were no stars below their horizons. And of course you know that wherever you are on earth, you look up into a sky where stars are visible at night (or would be, if there were no clouds in the way). The sky, then, is not just the half-sphere that we see. It must be, the early astronomers reasoned, a complete sphere (Fig. 1-4) encircling the earth. They thought of it as a hollow sphere of clear crystal on which the stars are pasted like so many jewels. They called it the *celestial sphere.*

The Silent Stars

If you could communicate tonight by telephone or ham radio with others looking at the sky at the same time from many widely scattered places on the earth, you would find that each of these persons had his zenith at a different place among the stars (Fig. 1-5). Therefore, the horizon line of each (90° down from the zenith) would also lie at a different place among the stars; it would cut the celestial sphere along a different line. The visible half of the celestial sphere would not be identical for any two of these people, although two near each other—one in New York City and one in Philadelphia, for instance— would see almost the same stars. An observer at the North Pole would see the Dippers, Draco, Cassiopeia, and Cepheus surrounding his zenith. Observers in Canada and the United States would

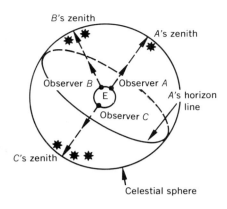

Fig. 1-5 The position of an observer's zenith (and his horizon) among the stars on the celestial sphere varies with his location on the earth. Stars near the zenith for observer A are near the horizon for observer B (and vice versa), and are invisible to observer C.

see these constellations successively lower in the sky. At the equator some of these five constellations are below the horizon. As you travel south more of them are hidden and other stars appear that are never seen in the northern hemisphere.

By putting together observations from all over the earth, the individual stars and the constellations can be drawn on the celestial sphere just as the oceans and continents, mountains and rivers, and cities and towns can be plotted on a globe of the earth. The stars and constellations thus become "skymarks" in the same way that features on the earth are landmarks.

About A.D. 150, an astronomer named Claudius Ptolemy lived in Alexandria, Egypt. Ptolemy summarized the many conclusions from observations of the sky like those you are making. He recorded them in a book. This book won so much admiration that it became known as *The Almagest,* a Greek-Arabic term meaning "The Greatest [of books]." He concluded that the earth, at the center of the "sphere of the stars," is very small—a mere dot in comparison to that outer sphere. If it were not, he pointed out, the horizon would divide the celestial sphere into two parts of unequal size. In Figure 1-5, the earth and the starry sphere had to be drawn close in size in order to fit the drawing on the page. And, as you can see, the horizon does divide the celestial sphere unequally. But, as Ptolemy saw, maps of the stars show that the horizon divides the celestial sphere exactly in half.

Ptolemy also noticed that each constellation appears equal and similar from wherever on the earth one views it. If the celestial sphere were not almost infinitely far away from the earth, constellation ✴ ✴ and its stars (Fig. 1–5) would look larger to observer B than to observer A. Cassiopeia, for instance, would look larger or smaller, depending on where you were when you saw it. Also, if the celestial sphere were not at an immense distance from the earth, the relation of stars to each other would appear to change, depending on the angle you saw them from. Yet the "W" of Cassiopeia has the same shape for those viewing it near the zenith as for those people who see it near the horizon. (Fill three or four balloons with helium and let them come to rest on the ceiling. Then walk about the room, looking up at them, and you will see how their pattern changes.)

From these arguments, and others, Ptolemy concluded that the radius of the celestial sphere must be very much larger than the earth's radius. This is another way of saying that the stars are at immense distances from us. This must also mean that the stars are very bright objects, since we can see them at such great distances, even as tiny points of light.

It may surprise you to learn that Ptolemy also concluded that the earth is round—or, more correctly stated, that it is a sphere. If we agree that the stars are at extremely great distances, we would expect the position of the zenith to be near the same star for all persons on a flat earth. If the earth were flat, straight up would be the same direction for every-body, and all the ham radio operators you contacted would look up and see the same star overhead. But straight up is not the same direction for everyone. The man in Mexico sees his zenith south of the star seen overhead in St. Louis. And the man in Canada sees his zenith to the north of that star (Fig. 1-6).

But a curved earth is not necessarily a spherical earth. A sphere has the same curvature all over its surface. At this point, we can follow only Ptolemy's argument that the earth is curved uniformly from north to south. You probably see Polaris about 30° to 40° above the northern horizon (at an *altitude* of 30° to 40°).

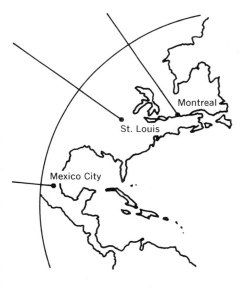

Fig. 1-6

At the North Pole (latitude 90°) you would see it almost in the zenith (almost at altitude 90°). At the equator (latitude 0°), Polaris is almost on the horizon (altitude 0°). As you travel northward or southward, the altitude of Polaris changes by the same amount that your latitude changes—1° each 69 miles. This is possible only if the curvature of the earth is uniform from north to south, as it is on a sphere.

Additional Reading

MARTIN, M. E., *The Friendly Stars* (rev. by D. H. Menzel): New York, Dover Publications, Inc., 1964.

PAGE, L. W., *A Dipper Full of Stars:* Chicago, Follett Publishing Company, 1964, pp. 13-114.

| **The Changing Sky**

After you have found the Big Dipper and Cassiopeia in the sky, go out once more after two or three hours to make sure that you can still recognize them easily. You will be surprised to see that these constellations are no longer in quite the same locations. The Big Dipper is still a dipper and Cassiopeia is still a "W." You can still draw a straight line from the end star of the Big Dipper's handle, through Polaris, to the end star in Cassiopeia (Fig. 1-1). Polaris is still at the same distance from each end of this line. However, both star groups have swung around relative to the zenith and to the horizon. If you were familiar with the entire sky and knew all the constellations, you would see that all of them had moved, even though their shapes and distances from each other are the same.

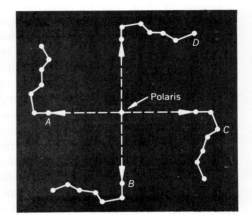

Fig. 2-1

What happened during the time that you were inside? If you had been watching the sky carefully, you would have seen the Big Dipper move slowly and continuously. It moved counterclockwise (opposite to the direction in which the hands of a clock move) about a point in the sky. And, like a clock hand, it moved with a circular motion. If you watched it through the night, and made a drawing of it soon after darkness fell and another drawing six hours later, your drawings would be as different as two successive ones in Figure 2-1. At daylight the Dipper's stars get drowned in the blue sky. But when darkness falls the next evening, the Dipper is back in the sky where you first sketched it.

The hour hand on a clock moves in a circle around the clock face once every 12 hours. The line joining the two outer stars of the Big Dipper's bowl and pointing to Polaris (Fig. 2-1) is like a clock hand. But this clock hand completes its circle around a small part of the sky once every 24 hours. Clock hands go around in a circle with its center in the middle of the clock face, and they always line up with this center. Carrying the two Dipper stars with it, the line we have drawn in the sky moves in a circle whose center seems to be Polaris. Each of these stars describes a circle around Polaris, and the one farther out completes its larger circle in the same amount of time as the star nearer Polaris. At the center, Polaris seems to be stationary, just like the axle carrying the clock hands, or the hub of a turning wheel.

If you were to make more careful measurements, you would see, as the early astronomers did, that even Polaris moves. It goes around a tiny circle (less than

1° in radius) every 24 hours. The center of this small circle is also the center of the larger circles followed by the stars of the Big Dipper. And it is the center of the even larger circles followed by the stars of Cassiopeia.

What about the other stars in the sky? If you follow their motions through the night, you will see that each one moves in a circle around this same point in the sky. Each one takes 24 hours to complete its circle. Figure 2-2a is a time-exposure photograph of the stars near Polaris. The photograph was made with the shutter of the camera open for two hours. The trails show how each star moved during that time.

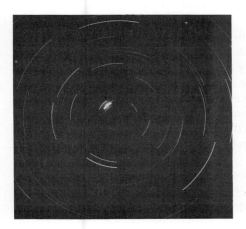 

Fig. 2-2 (a)Time-exposure photograph of the stars near the north celestial pole during two hours. The short arc very near the center was made by Polaris. (b)Time-exposure photograph of stars about 90° from the north celestial pole, near the celestial equator. [Yerkes Observatory photographs.]

This point in the sky, around which you see the stars move, is called the *north celestial pole*; and its direction is by definition true north. Stars farther and farther from it move in larger and larger circles. Time-exposure photographs of stars 45° or more from Polaris look like those in Figure 2-2b. The curves of these huge circles are so gentle that the stars appear merely to be moving from east to west across the sky.

Throughout all this rotation, the outlines of the constellations do not change. The stars remain at the same distances from one another, because each one moves around the pole through the same angle — the same fraction of a circle — in the same length of time. They complete their circles in 24 hours. How far do they go in one hour? In one minute?

If the north pole of your sky were at the zenith, as it is seen from the North Pole of the earth, no stars would rise or set. The same ones would always be above the horizon, continually circling the north pole of the sky. But you don't see the north celestial pole at your zenith. If you live in the United States it is, at most, halfway up from your northern horizon. You will recall from Chapter 1 that the altitude of Polaris is equal to your latitude on the earth (both measured in degrees). This is, of course, if you are observing at sea level. From elevations above sea level, the horizon line is lower, and therefore the altitude of Polaris is a little larger.

If you think about it a minute, you will see that only the stars whose distances from the north celestial pole are less than your latitude will stay above the horizon all around their circles. The other stars dip below the horizon during part of each 24-hour period. Unlike the constellations near Polaris (the Dippers, Draco, Cassiopeia, and Cepheus), these stars rise in the east, move across the sky, and set in the west. The farther they are from the north celestial pole, the shorter is the time that they are in your sky.

"Up the dome of heaven, like a great hill / I watch them marching, stately and still," wrote the poet, Sara Teasdale. The ancient astronomers were watching the stars intently, too, and they had a theory to explain their observations. Ptolemy recorded this theory of the universe in *The Almagest*. The whole celestial sphere, he said, carries the stars fixed to its surface and rotates around us once every 24 hours. This rotation carries the stars upward in the east, downward in the west (Fig. 2-3a). Its rotation is around an axis that passes through the north celestial pole and through the south celestial pole, which is 180° away and hidden from our view below the southern horizon.

If you consider it carefully and really observe the motion of the sky, removing from your mind all the things you have merely been told are so, you will see that this is an obvious and valid conclusion to draw from the daily motion of the stars.

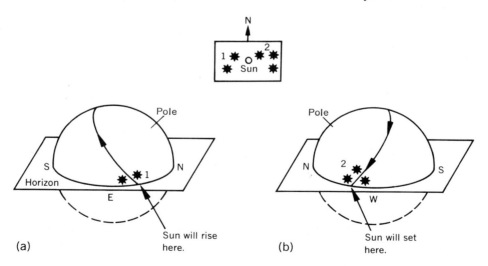

Fig. 2-3 (a) Constellation 1 rises in the east, followed by the sun, as the celestial sphere turns westward. (b) As the sun sinks below the western horizon, constellation 2 is seen for a few minutes just above the horizon. Then the westward turning of the celestial sphere carries it out of sight too.

14

It is pleasant to think of ourselves at the center of the universe, apparently the ones for whom this attractive and changing celestial display was planned. Nothing that you see, or feel, or hear, suggests that the earth has any motion of its own.

But if we are to accept this theory of the universe, it must explain all the observations that we can make. What about the sun? How does it fit into the picture? At first glance, this seems simple. The sun, a star much larger than the others, shares the daily westward motion of the celestial sphere. If you could see the stars in the daytime, you would see the constellations accompanying the sun. As you watch the sun go down and the stars come out in the darkened sky, you can see which constellation is near the horizon at the place where the sun disappeared. Soon this constellation, carried by the westward rotation of the celestial sphere, follows the setting sun and sinks below the horizon (Fig. 2-3b). Then, next morning just before dawn, you would see another constellation coming up in the eastern sky. As these stars rise higher in the sky, you see the sun also rise at the same place on the horizon and drown them out in its strong light.

Early astronomers reasoned that between sunrise and sunset, the sun journeys across the sky among the stars that lie between these two constellations. The stars are still there when the bright sun is in the sky, but it blots them out. Ptolemy concluded that as the celestial sphere turns, it carries the sun and the stars up from the eastern horizon and across the sky. As the sphere turns farther westward, the sun sinks out of sight below the western horizon along with the other stars far from the north pole of the sky.

If you look at the sky at dawn and dusk from time to time, however, you will see that different constellations appear with the sun at dawn and at sunset after two or three weeks. The sun does not travel across the sky with the same stars day after day. The early astronomers saw that, unlike the stars, it does not stay fixed in place on the turning celestial sphere.

Using the same laboratory and the same kind of equipment that you have — the sky and their eyes — these pioneer scientists made more observations. They kept track of which stars were near the sun and found the reason for their being there: The sun keeps slipping gradually eastward among the stars. In the course of about 365 days it completes one full trip around the sphere of the sky, back to the same place among the stars. This yearly trip is always along the same path, crossing 12 constellations that the early astronomers called the *zodiac*, a Greek word meaning "circle of animals." This is appropriate because some of the constellations of the zodiac represent animals. Each July 10, the sun enters the constellation Cancer; on August 10 it enters Leo, on September 10, Virgo, and so on through the year.

Two motions of the sun must be explained: a *daily* westward motion (explained by the turning of the whole celestial sphere) and a *yearly* eastward motion. It seemed obvious to the early astronomers that the sun crawls eastward along a set path, the *ecliptic,* over the surface of the sphere (Fig. 2-4). They assumed that the sun must be much larger than the other stars since it appears so much larger. In a second explanation, the sun was pictured as fastened to a sort of moving ring, the ecliptic, much closer than the stars on the celestial sphere. In this case, the stars were believed to form a background far behind the sun, and hence the sun need not be bigger or brighter than they. The sun's ring turns slowly eastward as the celestial sphere turns westward once a day, carrying the sun with it. The sun's ring, however, takes a year for a complete turn.

A theory should never be more complicated than needed to explain the observations. As far as our observations go, both of these explanations are good theories of the sun's yearly eastward motion among the stars, and we can choose the one we prefer. They also explain other observations that you have made: the changing seasons and lengths of days.

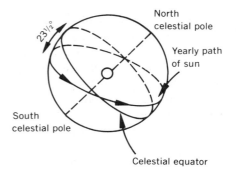

Fig. 2-4 The celestial sphere showing the ecliptic, the yearly path of the sun against the background of the stars, and its relation to the celestial equator. The constellations of the zodiac lie in a belt extending about 8° on either side of the ecliptic.

When the ecliptic, or yearly path of the sun, is plotted among the stars on the celestial sphere, it is found to be a circle inclined 23½° to the celestial equator (Fig. 2-4). In the northern hemisphere of the earth, the sun is 23½° above the celestial equator (the nearest it gets to the north celestial pole) about June 22. On that day the sun is visible longer than any other day of the year — as you would expect. When it is 23½° below the celestial equator (the farthest it gets from the north celestial pole) about December 22, it is above the horizon for a much shorter period each day. Between December 22 and June 22, the days gradually become longer and the nights shorter; then, after June 22, the days gradually become shorter and the nights longer.

The varying daily periods of sunlight and, more important, the changing angle of the sunlight on the earth, bring about the changing seasons. In winter, when the sun is lower in the sky, its rays strike us from a lower angle and hence are spread over a larger area (Fig. 2-5).

It is time to go out again to look at the sky. On summer evenings you may have noticed a conspicuous group of bright stars at the top of the sky. During the following winter, wishing to show off how well you know the sky, and forgetting the sun's eastward drift among the stars, you might try to point out this star group to one of your friends. Of course you would find that it is not there, even though you are in the same place and it is the same hour that you used to see it. Looking more carefully at the sky, you notice that although some of the constellations you had seen on summer evenings are in the sky, they are in different places with respect to the zenith and the horizon. You also see some new constellations, like the conspicuous Orion, that were not in the summer sky at the same hour.

What happened? The early astronomers noticed this change in the evening sky throughout the year. How did their theory explain it? Like us, these people told time by the sun. Noon, to us as it was to them, is the moment when the sun is highest in the sky. It is noon for you when the sun in its daily journey from east to west crosses the *meridian*. (The meridian is the line on the celestial sphere from the north point of your horizon through the zenith to the south point.) Since the sun is slipping backward (eastward) among the stars 360° per year or about 1° each day, the stars are continually getting ahead of the sun. A star that crosses the meridian at noon today (when the sun does) will cross it four minutes before noon (four minutes ahead of the sun) to-morrow. Thus, if we could see the stars at noon, the sky would look the same at noon today as it did at 12:04 yesterday.

The sky at 9:00 tonight will look to you just as it did at 9:04 last night. By the time 30 days have gone by, the star will be 120 minutes, or two hours, ahead of the sun. As a result, the sky on December 1 at 9:00 P.M. will present the same appear-ance as it did on November 1 at 11:00 P.M. Similarly, on June 1 at 10:00

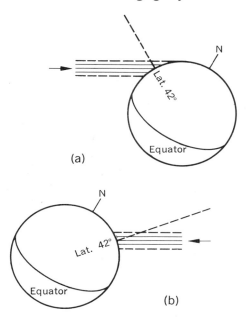

Fig. 2-5 When sunlight strikes the earth at a low angle as on December 22 (a), it is spread out more than when it comes in nearly vertical on June 22 (b).

P.M., the same stars will be in the sky as on December 1 at 10:00 in the morning, and in the same positions. This is why magazines like *Sky and Telescope* and *Natural History* carry monthly evening sky maps, and why the summer sky (Fig. 2-6a) holds different constellations than the autumn sky (Fig. 2-6b), the winter sky (Fig. 2-6c), and the spring sky (Fig. 2-6d).

That is why, of course, we had to tell you in Chapter 1 to find the Big Dipper and then twist the book so that the Dipper was at the same angle that you saw it in the sky. There was no way to tell on what date or at what hour you were going to start reading this book. Figure 2-1 at the beginning of this chapter shows the change in position of the Big Dipper throughout the night. It also shows the Dip-per's change in position at the same hour throughout the year.

Additional Reading

ASHFORD, T. A., *From Atoms to Stars:* New York, Holt, Rinehart and Winston, 1960, pp. 32-39.

PTOLEMY, CLAUDIUS, "The Heavens Rotate as a Sphere" in *A Source Book in Greek Science* (M. R. Cohen and I. E. Drabkin, eds.): New York, McGraw-Hill, 1948.

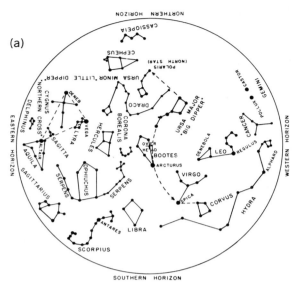

THE NIGHT SKY IN JUNE

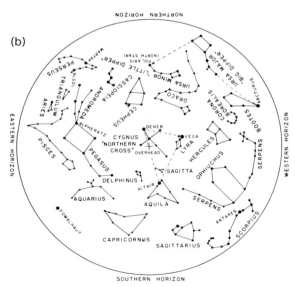

THE NIGHT SKY IN SEPTEMBER

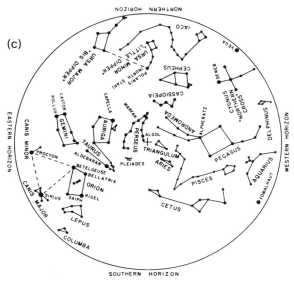

(c)

THE NIGHT SKY IN DECEMBER

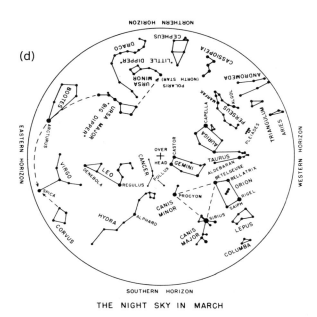

(d)

THE NIGHT SKY IN MARCH

Fig. 2-6 [From C. H. Cleminshaw, Griffith Observatory and Planetarium, Los Angeles.]

Using one of the star maps in the preceding chapter, you may go out one evening to add another constellation to those you know. Perhaps you find the sky so brightly lit by the full moon that stargazing is impossible. This is just the time to do some moongazing instead. The moon, in the southeastern part of the sky, is steadily moving westward. If you watched it all night, you would see it set toward dawn. Tomorrow night you will see it rise again over the eastern horizon—but almost an hour later. Like the sun, it appears to share in the daily westward turning of the celestial sphere, although it lags a bit.

Watch it for two weeks, though, and you will see several differences between it and the sun. Most striking of these, of course, are the moon's changes in appearance. You will see it go from full moon, to three-quarters moon, to half moon, to crescent, waning until no moon is visible at all. In the next two weeks you will see these phases in reverse order, as the moon is waxing.

You have found that the moon rises and sets about 50 minutes later each day than it did the day before. Since we tell time by the sun, this suggests that the moon is slipping eastward among the stars faster than the sun is. Because even the brightest moon doesn't completely drown the light of the stars, it is possible to see with what constellation the moon crosses the sky each night. And when these are recorded, it is evident that the moon is, indeed, slipping backward (eastward) on a path passing through the 12 constellations of the zodiac. Its rate is about 13° per day, while that of the sun is only 1° per day. The result is that the moon completes a circle around the celestial sphere and comes back to the same place against the star background once each month, while the sun takes a year to complete its circuit. (Just as the sun determined the length of the year, so the

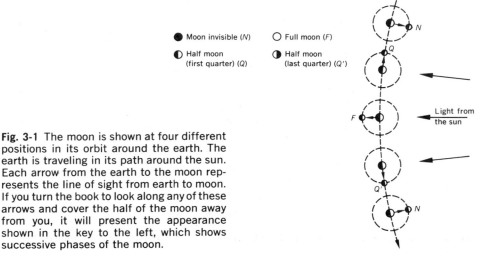

Fig. 3-1 The moon is shown at four different positions in its orbit around the earth. The earth is traveling in its path around the sun. Each arrow from the earth to the moon represents the line of sight from earth to moon. If you turn the book to look along any of these arrows and cover the half of the moon away from you, it will present the appearance shown in the key to the left, which shows successive phases of the moon.

moon determined the length of the month. However, the 12 months we have during the year are reckoned a bit too long, so the moon is not full on the same date each month. The true month is 29¼ days, not 30 or 31.)

Your records will also show that the full moon always rises at about sunset and sets about dawn. This means that the full moon is always on the opposite side of the earth from the sun. The thin crescent seen just before and just after "dark of the moon," rises in the early morning just before the sun or sets in the early evening just after the sun. The moon is then on the same side of the celestial sphere as the sun. When the lighted portion of the moon is growing larger (waxing), the moon rises during the daytime hours. You can often see it in the sky with the sun. In its waning phases the moon rises during the nighttime hours.

The positions of the sun and the moon at each moon phase are a clue that early astronomers interpreted in the same way we do today: When the moon is on the opposite side of the earth from the sun, the sun lights up the whole surface of the moon facing us. When the moon comes between us and the sun, the far side of the moon that we cannot see is lighted. Between these two extremes, we see less and less of the moon's lit surface, and then more and more of it (Fig. 3-1).

This explanation points out another difference of the moon: Unlike the sun and the stars, it does not give out light of its own; it shines only by reflected light. It sends secondhand sunlight back to us. When the moon is a thin crescent, the part of its face that is not illuminated by the sun is not completely dark (Fig. 3-2). It glows faintly, reflecting light from the sunlit side of the earth that faces it.

Its phases showed the ancients that the moon could not be crawling eastward along the surface of the celestial sphere because then it couldn't get between us and the sun. It must, they reasoned, be nearer us than the sun. They thought of the moon as attached to another moving ring, attached somehow to the sphere, and located inside the ring bearing the sun. The ring carried the moon eastward around the earth once each month, while the rotation of the sphere carried it westward across the sky once each day. It was clear to these observers that the moon must be smaller than the sun; although the moon is nearer to us, it appears almost the same size (about one-half of a degree in diameter) in the sky.

Although eclipses of the sun and moon filled the minds of most people with superstitious fear, the early astronomers welcomed them as evidence for their theory. They saw that all solar eclipses take place in the dark of the moon and all

Fig. 3-2 "The old moon in the new moon's arms." [Yerkes Observatory photograph.]

lunar eclipses at full moon. Of course, if the moon gets directly between us and the sun, it can eclipse the sun (Fig. 3-3a), since both have about the same angular size. And when the moon and the sun are lined up on opposite sides of the earth, the shadow cast by the earth can darken the moon (Fig. 3-3b).

Why, then, aren't there two eclipses each month, one of the sun and one of the moon? There would be if, as the ancients put it, their two "rings" were inclined at the same angle—if their two paths among the stars were identical, both along the ecliptic. For then the moon, sliding eastward faster than the sun, would have to cover the sun each time it passed by. Careful measurements show, however, that the path of the moon is tipped 5° to the path of the sun. The moon is usually slightly above the sun (north) or slightly below (south) as it passes the sun. Only where the moon's path crosses the sun's path (as seen against the background of the stars) can there be an eclipse. It will be a solar eclipse if the sun happens to be there, and a lunar eclipse if the sun happens to be on the opposite side of the sky.

A good theory must do more than just explain a few observations. The early astronomers found that, knowing the positions of the sun and the moon, the path of each, and the angular speed with which each was moving eastward along its path, they could predict successfully when eclipses would occur. It was through just such predictions that they became known as "wise men."

These early astronomers also used an eclipse to mark a definite time without instantaneous communication. They could not determine (as we can by radio or

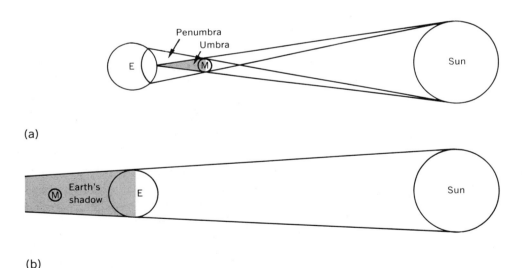

(a)

(b)

Fig. 3-3 (a) Conditions causing an eclipse of the sun. A total eclipse can be seen when the *umbra,* or real shadow, of the moon passes overhead. This shadow never exceeds 170 miles in diameter. The *penumbra,* the area surrounding the umbra, does not completely hide the sun, and observers in that lightly shaded area will see only a partial eclipse. Unlike the figure, the penumbra is a gradual shift from total dark to total light. (b) Conditions causing an eclipse of the moon.

telephone) where the sun was in the sky over Alexandria, Egypt, when it was on the meridian over Constantinople. But they realized that an eclipse was an event which took place at one moment in time. Ptolemy asked astronomers in many different places where the sun had been with respect to the meridian of each when each one had seen it eclipsed. Since each man's meridian passes through his own zenith, the different angles between the eclipsed sun and each of the meridians showed the angles between their zeniths. These differences (angles) turned out to be proportional to the observers' east-west distances from each other, 1° for each 69 miles. This could only be so (Fig. 3-4) if the east-west curvature of the earth were uniform. Thus, having shown by the changes in elevation of Polaris that the earth was uniformly curved by the same amount north-south, Ptolemy could say that the earth is a sphere. This view has been held by astronomers for 2000 years and by most other people

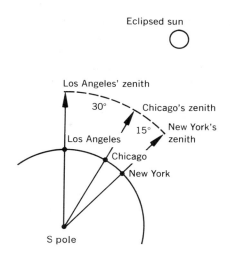

Fig. 3-4 When the sun is on the meridian at Chicago and it is noon there, the sun is 15° west of the meridian in New York City (1000 miles to the east) and it is 1:00 P.M. there. At the same moment, the sun is 30° east of the meridian in Los Angeles (2000 miles west of Chicago), and it is 10:00 A.M. there.

for 500 years. However, a look at the earth's curved shadow on the moon at the time of a lunar eclipse should have suggested it to them, as well as to astronomers.

As we considered the eastward paths of the sun and moon through the constellations of the zodiac, you may have looked for some of these star groups in the sky. Perhaps you became familiar with Leo the Lion (Fig. 3-5a), shining almost overhead on springtime evenings. The "sickle" which forms the forepart of the lion, with a bright star in the end of its handle, is easy to spot in the sky. Perhaps one summer evening when you went out to see Leo, now stalking along the southwestern horizon, you found the sickle looking like Figure 3-5b. That second bright star certainly wasn't there when you looked at the constellation a month or two earlier. Star charts like Figures 2-6a and 2-6d do not show it either. As the summer advanced, Leo set with the sun, and you didn't see him again until early the next spring. Then you might have found him with only one bright star again, looking once more like Figure 3-5a.

Or let us imagine a group of shepherds, long before Ptolemy's day, whiling away the long nights by watching the stars. Imagine that on a particular night they were looking at the zodiac constellation of Taurus the Bull. It had two bright stars and several fainter ones, forming a letter "V" lying on its side. After a few nights, one of the more careful stargazers noticed that the brightest star of Taurus was no longer in quite the same position relative to the other stars in the constel-

23

The Moon and the Wanderers

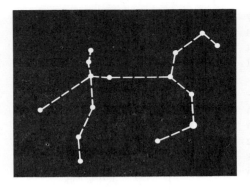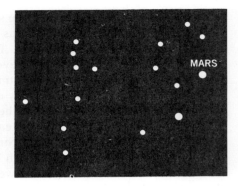

Fig. 3-5 (a) The constellation of Leo the Lion and (b) showing the appearance of the second bright star.

lation. His friends agreed that the constellation did look changed, and that none of the stars in Taurus except this one had moved with reference to the others. Several weeks of observation showed that this bright star in the Bull was slowly moving eastward among the stars. In time, its eastward path took it out of this constellation (which then looked as it does in Figure 2-6c) and into another zodiac constellation, Virgo.

By Ptolemy's time, observations of the sky showed that five stars move among the fixed stars. These were called "πλανήτης," which means "wanderers" and is pronounced about like "planets." Each one was given a name in honor of one of the gods: Mars (the planet pictured with Leo in Figure 3-5b), Jupiter, Saturn, Venus, and Mercury.

If you were to chart the motions of Mars, Jupiter, or Saturn over a period of time, you would find that they cross the sky each night from east to west in one of the constellations of the zodiac, and that, in general, they slide slowly eastward among the stars like the sun and moon do. But every now and then you would see one of these planets slowly make a loop among the stars. After moving eastward for a little more than two years (Mars) or a little more than a year (Jupiter or Saturn), it slows up, reverses its path, and proceeds westward for a while before resuming its eastward path.

A planet does not repeat its path accurately among the stars like the sun does each year and the moon each month. Nor does it always move relative to the stars with the same angular speed.

The most beautiful of the "wanderers" is Venus, exceeded in brightness only by the sun and moon. Because Venus is never more than 47° east or west of the sun, you never see it in the middle of the night. It sometimes shines for nearly three hours after sunset in the western sky, when it is called the Evening Star, or for about three hours before sunrise in the eastern sky, when it is called the Morning Star. The very early astronomers had two names for this planet, for when they saw it in the morning sky they did not realize that it was the same planet they had seen some six months earlier in the evening sky. Later they realized that Venus was moving west to east and then east to west on either side of the sun. Mercury, very difficult to see and never more than 27° from the sun, behaves in the same way, about two months elapsing between its appearance as a morning star and as an evening star.

Ptolemy thought about how these peculiar motions of the planets could be fitted into the theory of the universe and summarized his idea in *The Almagest.* He pictured, as you may now expect, separate bands or rings around the inside of the celestial sphere, one for each planet. Each planet, however, was not fastened directly to its turning ring, as the sun and stars were to theirs. Instead, the center of the smaller moving ring was attached to the main ring, with the planet on its rim (Fig. 3-6). The smaller ring or wheel carried the planet around, so that it was sometimes moving in a direction opposite to the movement of the big ring. The rings turned at different speeds. As Ptolemy explained, these movements could produce the looped motions seen in the sky, or the back-and-forth motions of Mercury and Venus.

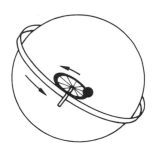

Fig. 3-6 Ptolemy's theory shows a planet moving along an epicycle, which in turn moves along a larger ring around the earth, thus producing the looped motions of the planets in the sky.

By Ptolemy's time, men had been accurately recording and timing the positions of the planets among the stars for several centuries. He estimated the ring sizes and speeds to explain all these observed and recorded motions. This explanation worked very well and was accepted for over a thousand years. The trouble was that the planets did not move among the stars exactly as the theory predicted. Ptolemy's successors would patch it up by adding more wheels or rings. A planet was supposed to be attached to a wheel whose axle was attached to another wheel, whose axle was attached to still another wheel, whose axle moved on the main ring. Before too long, as many as 18 wheels moving on each other at varying speeds were needed to explain the past and present positions of one planet in the sky.

Ptolemy's picture of the planets' motions was a mathematical model. He called the smaller rings or wheels *epicycles,* and probably never thought of them as material rings or wheels. He was using geometry to build up an explanation of observed movements, just as modern mathematicians use algebra. It is mathematically possible to represent any movements of the planets if a sufficient number of wheels within wheels are used.

By A.D. 1500, astronomers were using 40 epicycles to explain the past positions of Mars. The theory was becoming terribly complicated and, worse than that, it could only predict a planet's motion for a few years. It failed to explain all the observations. People accepted it for a long time because they believed that since a circle was the perfect form, motion in a circle was the perfect motion. Since they were dealing with celestial or divine objects, surely the only shapes and movements these objects could possibly have were circular ones. Hence the idea of rings and wheels within wheels, which all provided the appropriate circular motions. More important, there was nothing to suggest that the solid earth might be moving. Besides it was unthinkable that earth, the abode of man, could be anywhere but at the center of the universe.

But the difficulties of the Ptolemaic theory set the stage for a questioning of this whole interpretation of the universe. And, as we shall see, a new picture of our universe came from a different interpretation of the movements of the stars, sun, moon, and planets in the sky.

Additional Reading

> ASHFORD, T. A., *From Atoms to Stars:* New York, Holt, Rinehart and Winston, 1960, pp. 39-42.
> COLEMAN, J. A., *Early Theories of the Universe:* New York, New American Library (Signet Science Library), 1967, Chap. 7.
> TRICKER, R. A. R., *The Paths of the Planets:* New York, American Elsevier Publishing Company, 1967, Chap. III.

chapter 4 | # A Revolution in Revolutions

As far back as 300 B.C., Aristarchus, a Greek philosopher, pictured the sun at the center of the universe, with the earth moving around it. His idea did not cause much of a ripple in the scientific world. There was little support except for his estimate that the sun is larger than the earth. Every bit of evidence seemed against the possibility that the earth might be in motion. Moreover, at that time, the problem of accurately predicting the planets' paths or explaining thousands of careful observations of their positions in the sky had not yet arisen. An earth-centered universe was obvious, satisfying, and adequate.

So, until 400 years ago, the theory of Ptolemy held sway. For the final 1400 years of its reign, and especially throughout what are sometimes referred to as the Dark Ages, astronomers were kept busy patching up the epicycles to agree with the observations. In order to predict where a planet would be in the sky on a given date, they had to calculate the motion of as many as 40 of these wheels-within-wheels, each moving at a different speed. Even after all this tedious work, the predictions were not very accurate.

Then in 1543, the theory of a Polish monk was published under the title *First Account of the Revolutions of Nicholas Copernicus.* The revolutions were those of the earth and the other planets around the sun. But, as we think of it today, the title has a second meaning. The book began a revolution in ideas about the universe. The refreshing simplicity of Copernicus' universe contrasted with the evergrowing complexity of the Ptolemaic theory and helped to sell his idea (Fig. 4-1).

Copernicus, like all astronomers who came before him, believed that the stars are fixed to a celestial sphere and that the planets and the moon move in circles, each at unchanging speed. But, he said, the celestial sphere is motionless; it is the earth that is rotating daily. The sun is motionless too, and the planets revolve around it. In his theory, the earth is far from being the central body

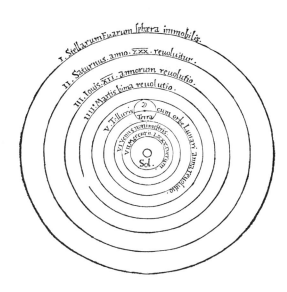

Fig. 4-1 The Copernican solar system, as pictured in a book published in 1566. [Yerkes Observatory photograph.]

of the universe. Instead, it is a planet, like the five "wanderers" that are seen only as small lights in the night sky. This demotion was hard to swallow, and it was even harder to believe that the solid earth could be in motion. What evidence did he give to back up his ideas?

First of all, Copernicus pointed out that a person moving at a constant speed may not be aware that he is moving if this motion requires no effort on his part. Have you ever looked from the window of a moving train and felt that the train was standing still while everything outside it was moving rapidly backward? Or when a car parked parallel to yours backed out, have you ever felt that your car was slipping forward? Motion of the observer and the observed cannot always be distinguished.

In the same way, Copernicus said, the daily westward motion of the celestial sphere, which appears to carry the stars, the sun, the moon, and the planets from east to west each day, is only an apparent motion. In reality, he said, the earth itself is rotating eastward, one complete rotation each day. The earth's rotation is around an axis (a straight line) passing through its North Pole and its South Pole. One end of this axis points toward the north celestial pole and the other toward the south celestial pole.

Copernicus then argued that the apparent annual motion of the sun was due to an annual motion of the earth in a circular orbit around the sun. As the earth moves in this orbit, we see the sun against a changing background of constellations.

A Revolution in Revolutions

Imagine a huge sheet of paper that passes through the earth at its equator and continues outward to cut the celestial sphere at its equator. This is the *plane* of the equator. The ecliptic (Fig. 2-4), as we have seen, is the observed yearly path of the sun against the background of the stars. Now imagine a second sheet of paper across the hollow interior of the celestial sphere, cutting it along the ecliptic. This is the plane of the ecliptic. The two planes intersect at an angle of 23½°.

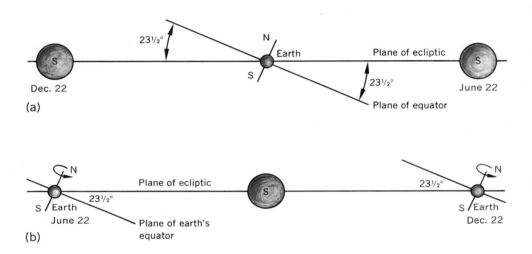

Fig. 4-2

To Ptolemy the sun's path around the earth lay on the plane of the ecliptic. Figure 4-2a shows a vertical slice through the two planes, with the sun at two positions on its path, December 22 and June 22. The sun, traveling around the earth, passes north of the earth's equator and then south of it. Copernicus, on the other hand, saw the setup as in Figure 4-2b. To him, the plane of the ecliptic contained the path of the earth around the sun. The earth is shown at two positions, June 22 and December 22. As the earth travels around the sun, the earth's equator is first above the sun and then below it.

Notice that the earth's equator always lies at 23½° to the plane of this yearly path, and that the earth's axis always makes an angle of 66½° to it. This diagram is undoubtedly a familiar one to you. Using this diagram and Figure 2-5, you can see why winter is colder than summer and why the northern hemisphere is shivering while the southern hemisphere is enjoying summer. Remembering that according to Copernicus the earth rotates on its axis once each day, we can learn more from this diagram. You can see why the days are longer than the nights in summer, why they are always of equal length at the equator (and everywhere else on March 21 and September 21), and why the North and South Poles have six months of darkness and then six months of continuous daylight.

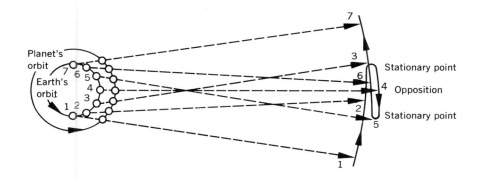

Fig. 4-3 How Copernicus explained the looping motions of Mars, Jupiter, and Saturn. As earth, moving on its orbit, passes a planet farther from the sun, our line of sight to that planet swings backward for a time, causing the apparent backward motion of the planet.

What you may not have thought of before is that these same seasonal effects could be produced by the movement of the sun along a path inclined $23\frac{1}{2}°$ to the earth's equator, together with a daily rotation of the celestial sphere.

So far, the theory of Copernicus does not seem to help much. It substituted movements of the earth — hard to believe — for motions of the celestial sphere and the sun. It told no more and no less about the causes of the seasons. And, as you probably have guessed already, it left the moon moving around the earth, for there is no other way to explain its phases, although Copernicus would illustrate them as we did in Figure 3-1.

However, if the earth is included among the planets, and if all six of them move around the sun in circles, the observed motions of the planets other than earth are more easily explained. Figure 4-3 shows that, viewed from the moving earth, a planet farther from the sun will appear to follow a loop at regular intervals each time the earth passes it. The planet, although generally moving eastward with respect to the starry background, will appear to stop, move westward for a while, and then resume its eastward course along the celestial sphere.

Figure 4-4 shows that a planet nearer the sun than the earth is will appear to move from one side of the sun to the other and back again as it travels around its orbit. Our earthly view of Mercury and Venus shows that they travel in smaller orbits around the sun than the earth does. Copernicus calculated the sizes of their orbits (relative to the earth's). When we see Venus farthest from the sun in the sky, the line from Venus to the sun makes a right angle with the line from Venus to the earth. The angle, seen from earth, between Venus and the sun is observed to be 47°. If we consider the earth-sun distance as 1.00, simple measurement shows the Venus-sun distance to be 0.72. In the same manner, the distance of Mercury from the sun can be determined to be 0.36 of the earth's distance. Circular orbits of the three planets can be drawn in the correct relative sizes.

Their looping motion on the opposite side of the sky from the sun shows that Mars, Jupiter, and Saturn move in orbits farther than earth from the sun. But how

29

much farther? And in what order from the sun do their orbits lie? As you can see from Figure 4-3, a loop of westward motion takes place when the earth overtakes and passes one of these planets. This passing occurs at regular intervals for each planet—for Jupiter it is every 1.09 years. But this is not the time that Jupiter takes to complete its circle around the sun (Jupiter's *period*). Nor is it earth's period (one year). But from these two bits of information Copernicus found the period of Jupiter.

We can see how he did it by an analogy. Consider the two hands of a clock, moving around a spot in the center of the clock face. The minute hand has a period of one hour; the hour hand has a period of 12 hours. At 12:00 the two hands lie along the same line to the center of the clock face. By 1:00 the minute

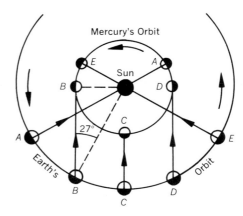

Fig. 4-4 Mercury and the earth in their orbits around the sun. Turn the book to look along the lines between the earth and Mercury, and you will see Mercury's changing position with respect to the sun. At positions A and E, Mercury is back of the sun and hidden by it. At position C, Mercury is directly in front of the sun and hence invisible. The orbit of Venus is about halfway between those of Mercury and the earth.

hand has completed one revolution; it is again pointing to 12:00. But it does not line up with the hour hand, because the hour hand has moved along one-twelfth of its circle in that hour. It is not until about 1:05 that the two hands line up again. Again the minute hand completes a revolution, but by then the hour hand has moved on, and they line up again at about 2:10. The interval between the lining up is about one hour and five minutes (1.09 hours).

If we compare the minute hand (period = one hour) to the earth (period = one year), and the interval between the lining up of the clock hands (1.09 hours) to the interval between Jupiter's loops (1.09 years), then Jupiter's period may be compared to the period of the hour hand (12 hours). So it should not surprise you that Copernicus' calculations showed Jupiter's period to be about 12 years. The period of each of the three outer planets may be worked out by similar reasoning because the intervals between the loops of each can be timed. Copernicus showed that Mars' period is a little over two years, and Saturn's about 30 years.

Next, Copernicus calculated the sizes of their orbits. Figure 4-5 shows the sort of observations that he used. It is a plot of the part of the celestial sphere on which the ecliptic and the celestial equator lie. The path of Mars is shown, and each dot is an observed position of the planet against the background of the stars at two-week intervals.

We will consider two positions of Mars in the sky, that of February 15 (M_1) as Mars was moving eastward and that of April 15 (M_2), taken midpoint on the loop as Mars appeared to move westward. These two observations were made two months apart. Therefore we know that the earth traversed one-sixth of its orbit (60°) in the meantime. In Figure 4-6a, the position of the earth when Mars ap-

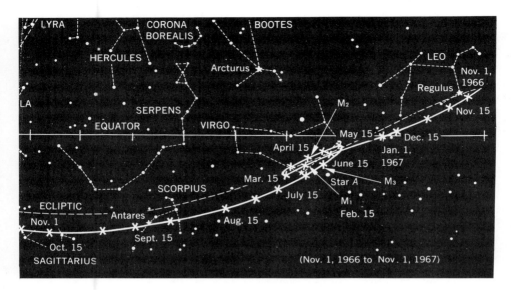

Fig. 4-5 The path of Mars (Nov. 1, 1966 to Nov. 1, 1967).

peared at M_1 can be plotted at E_1 and its position when Mars appeared at M_2 can be plotted at E_2.

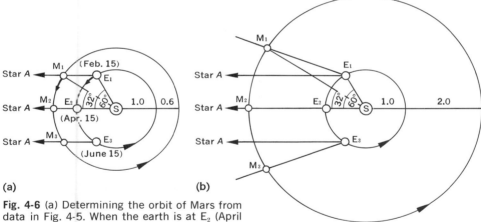

(a)

(b)

Fig. 4-6 (a) Determining the orbit of Mars from data in Fig. 4-5. When the earth is at E_2 (April 15), it is passing Mars because Mars is in mid-loop in the sky. Therefore, earth, sun, and Mars are lined up (SE_2M_2). Mars is seen near Star A in the sky, as in Fig. 4-5, so the line-up is SE_2M_2 Star A. Two months earlier, on February 15, the earth was 60° farther back on its orbit at E_1 and Mars was about 32° back on its orbit at M_1. But Mars was seen near Star A then, too, so the line-up was E_1M_1 Star A (parallel to SE_2M_2 Star A, because Star A is so far away). On June 15 the earth is at E_3 and Mars at M_3, again near Star A in Fig. 4-5, so the line-up E_3M_3 Star A is again parallel to SE_2M_2 Star A. When Mars' orbit is drawn to fit these observed directions, it must be 1.6 times larger than the earth's orbit. (The ratio is more accurately 1.52 as given in Table 4.) (b) If you draw Mars' orbit three times larger than the earth's, the observed directions E_1M_1, E_2M_2, and E_3M_3 do not come out lined up with Star A.

31

Because Mars' period is a little less than twice the earth's, it moved only a little more than 30° around the sun while the earth moved 60° around a smaller orbit. Because at E_1 and E_2 we saw Mars at the same position against the stars (which we will consider infinitely far away), the lines from the earth to Mars at these two times must be parallel. The job in calculating an orbit for Mars is to satisfy these two conditions. Figure 4-6a shows that orbit, and it places Mars 1.6 times farther from the sun than the earth is. Of course, all the other observations of Mars shown in Fig. 4-5 must fit this orbit too. In Fig. 4-6b, where Mars' orbit is drawn three times as large as the earth's, the line E_1M_1 is not parallel to the E_2M_2 line. You can try other orbit sizes and they will not fit the observations.

In a similar way, the orbit of Jupiter can be calculated (Fig. 4-7), using the observed positions of Figure 4-8 and remembering that Jupiter's period is 12 years. The orbit of Jupiter turns out to be five times as large as the earth's.

Copernicus worked almost a lifetime with very detailed observations covering hundreds of years of recorded planet positions. He established the scale (Table 1) and the pattern of his "solar system," and showed that the nearer a planet is to the sun, the faster it moves in its smaller orbit. The distance around a circular orbit is $2\pi R$ (where R is the radius, or distance of the planet from the sun). The speed of a planet in its orbit is then $2\pi R / P$ (where P is the period of the planet). If we compare the relative speeds for three planets, we can see this pattern. For earth it is $2\pi 1 / 1$, for Mars $2\pi(1.5)/2$, for Jupiter $2\pi 5/12$. A little arithmetic

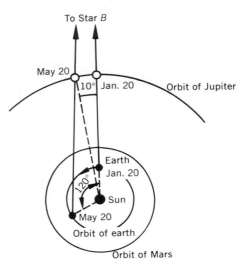

Fig. 4-7 Calculations similar to those used to determine the orbit of Mars (Fig. 4-6) can be used to calculate the orbit of Jupiter.

will show that while the earth moves a mile along its orbit, Mars moves 0.75 mile, and Jupiter 0.4 mile.

In general, then, Copernicus' system could explain the observed motions of the sun, moon, and planets. However, in detail it was a little off the mark. He had centuries of observations (many of them accurate to almost one-tenth of a degree and timed to an hour or so — $1/9000$ of a year) that did not fit his theory exactly. He was constantly making predictions of planets' positions and finding that these predictions were off slightly, perhaps only a degree or two. To make them fit precisely, he had to add 34 epicycles to his system: four to the orbit of the moon, three to the earth's orbit, seven for Mercury, and five each for the other planets. Then the observations fit. This was a great simplification of Ptolemy's universe which by then needed about 300 epicycles.

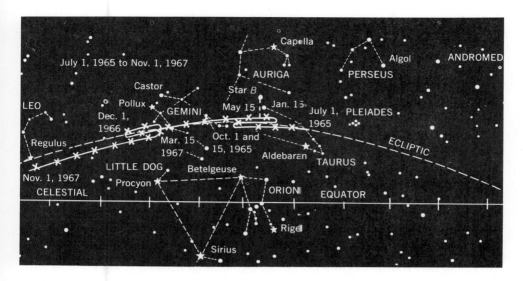

Fig. 4-8 The path of Jupiter (July 1, 1965 to November 1, 1967).

Planet	Determined by Copernicus	Modern determinations
Mercury	0.36	0.387
Venus	0.72	0.723
Earth	1.00	1.00
Mars	1.5	1.52
Jupiter	5	5.20
Saturn	9	9.54

Table 1 Relative distances of planets from the sun.

Additional Reading

DINGLE, HERBERT, "Copernicus and His Predecessors" in *Astronomy* (Samuel Rapport and Helen Wright eds.): New York, New York University Press, 1964.

KOESTLER, ARTHUR, *The Sleepwalkers* (Part 3: "The Timid Canon"): New York, Grosset and Dunlap (The Universal Library), 1963.

TRICKER, R. A. R., *The Paths of the Planets:* New York, American Elsevier Publishing Company, 1967, Chaps. VII, VIII, and IX.

| # Tycho, The Observer

Copernicus did not prove that the earth rotates and revolves around the sun. He firmly believed that it did, and astronomers had to agree that his theory explained observed movements in the sky equally as well as the accepted one. Indeed, they soon found that tables of the future positions of the planets, based on his theory, were just as accurate as those calculated with much more difficulty by the Ptolemaic method.

Copernicus' theory caused a tremendous stir. Almost everybody was against it. It seemed an affront to common sense. (Even today, when you "know" that the earth rotates and revolves, it is hard to imagine it whirling like a top as it rushes through space.) If the earth were rotating, people asked, wouldn't everthing near the equator go flying off? Copernicus had foreseen that objection. His rebuttal had been that the celestial equator would be going even faster than the earth's equator so that the stars could get all the way around their large circles each day (just as a point on the rim of a wheel moves through a greater distance each minute than the hub does). He pointed out that there would be even greater danger of the celestial sphere flying apart.

The scholars said that this was nonsense. Hadn't Aristotle, most revered of the Greek philosophers, said that the celestial objects — stars, sun, moon, and planets — were quite different from earthly objects? Hadn't he clearly stated that the natural laws applying to earth and the things on it do not apply to the "divine" objects? Of course the celestial sphere could rotate as fast as it wanted to, undisturbed! But the earth could not. Besides, it was like social climbing to class the dirty old earth as one of the divine celestial objects.

The clergy liked Copernicus' theory even less, but for the opposite reason. It was wicked, they said, to think that the abode of man, obviously favored by heaven, might not be the center of the universe. It was sinful to think of it chasing ignominiously around the sun, in the company of the "wanderers." Besides, why would Joshua have commanded the sun to stand still if it were not moving?

Many astronomers also opposed the new theory, even while welcoming its simple accuracy in determining future positions of the planets. This was important to them because at that time astrology (the telling of fortunes by the positions of the sun, moon, and planets) was tremendously popular. Almost everyone who could afford it consulted an astrologer, and many a king and nobleman hired one to work for him alone. The astrologer had to know the exact position in the zodiac of each planet at future dates and hours so that he could advise his employer when to wage wars, sign treaties, or be especially wary of enemies. Many of these astrologers were really astronomers, some working with tongue in cheek, who used the equipment and leisure provided by their employers to carry on astronomical studies.

Among the astronomers who did not accept the Copernican theory was a Danish nobleman named Tycho Brahe, born in 1546, three years after Copernicus'

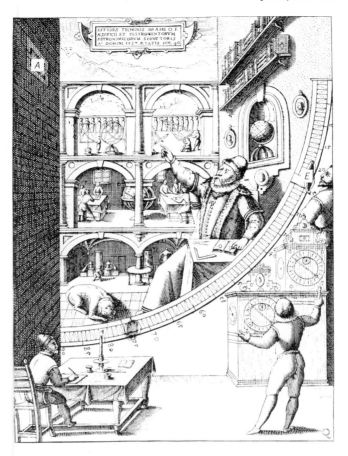

Fig. 5-1 [The Granger Collection.]

death. He was employed by the King of Denmark, who built him an elaborate astronomical observatory in a comfortable island castle. The many instruments in this observatory were more precise than any used before by astronomers. Tycho designed most of them and supervised their construction.

One of these instruments was a "mural quadrant," shown in Figure 5-1. The center background of the picture is the wall against which the quadrant (quarter circle) was set. This wall was decorated with a painting of Tycho, dressed for a winter's night of observing. Other instruments and activities in his observatory are also shown in the picture. The quadrant had a 6-foot, 9-inch radius and was marked with numbers outside the rim, showing angles above the horizon from 0° to 90°. Tycho is shown observing, at the extreme right. By sitting down low (behind the clocks) or climbing up steps (not shown) on the right, he could sight a star, planet, or comet through the window *A* at the left. This window faced south, where a faithful astrologer at a northern latitude would be looking at the zodiac. He sighted on an object in the sky, lining it up with the intersection of two wires marking the center of the window. Then he moved the sliding sight *E* on the quadrant until it, too, lined up, and read the angle from the quadrant, measuring between the degree marks along the zigzag lines.

For 20 years Tycho spent every clear night observing the sky. He was the most skilled observational astronomer that the world had seen, and one of the most illustrious of all time. Most of his observations — all made without a telescope — have proved to be accurate to one-sixtieth of a degree (one minute of arc).

In 1572 he saw a new star appear in the constellation Cassiopeia. It grew in brightness until it rivaled Venus. He watched it for 16 months until, after gradually becoming dimmer, it finally disappeared. His most careful measurements showed not the slightest change in its position relative to the other stars. Therefore, he knew that it belonged on the celestial sphere. His faith in Aristotle, who had pronounced the heavens perfect and unchangeable, was shaken.

Five years later, a brilliant comet like the one in Figure 5-2 appeared in the sky. Aristotle had taught that comets were something abnormal in the earth's

Fig. 5-2 Photograph of a 1965 comet, Comet Ikeya-Seki, photographed by William Liller on Oct. 29. The lights of Pasadena, California, can be seen in the foreground of the picture.

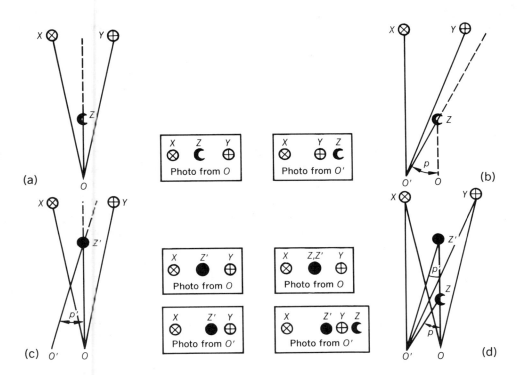

Fig. 5-3 (a) Three objects X, Y, and Z appear to an observer at O to be equidistant from each other, as shown in the "photo." The observer's line of sight to Z is the line OZ. (b) If O moves to O', Z seems to change its relation to X and Y, as shown in the "photo" from O'. The line of sight $O'Z$ differs from the line OZ by the angle $O'ZO$. This angle is the parallax of Z as viewed from O and O'. The observer would see the *same* effect if X, Y, and Z keep their same formation and move to the right through the distance OO' while the observer stays at O. (c) Suppose that Z is at Z', twice as far from the observer at O. Then X, Y, and Z' are in the same relation to each other on a "photo" from O as are X, Y, and Z in the "photo" in (a). If O now moves to O', Z' seems to change its relation to Y and Z, but the change is only half as much as it was for Z. (d) The angle $O'Z'O$ is smaller than the angle $O'ZO$. The parallax of Z' is smaller than that of Z. The parallax of X and Y is even smaller, and if we move them farther and farther from O, it becomes less and less until, when they are at an infinite distance from O, it becomes zero. The lines of sight of all observers to X would then be parallel, and the lines of sight of all observers to Y would be parallel. If we think of X and Y as two stars and Z as a comet or planet, the parallax of Z can be determined by measuring the angular distance that it moves in relation to X to Y, as in the "photo" insets.

atmosphere. Tycho, however, found that the comet was not a small object near the earth, but a large object much farther away than the moon. He reached this conclusion by trying to measure the comet's parallax.

Parallax is the change of position of an object against a distant background as its observer moves sideways. It is the difference in direction of the lines of sight to that object from two different places (Fig. 5-3). The nearer the object is, the larger is its parallax angle as viewed from two places. The same effect would be produced if the three objects in Figure 5-3, keeping their same formation, moved while the observer stood still. (Hold up a pencil at arm's length. View it first with one eye, then the other. Repeat this at one-half arm's length.)

If the distance between two observers is known, the distance to the object can be determined (Fig. 5-3d). We know that the parallax of the moon is larger than that of a more distant object, such as Tycho's comet, as determined from two simultaneous observations at different locations.

With the poor communication of that time, and isolated on his island in northern Europe, Tycho couldn't compare his line of sight to the moon, or to his comet, with sights of other observers at the same moment. Instead, he compared his own lines of sight to these objects at different times. At first glance, this seems simple. Both the background stars and the moon were being carried through the same angle per hour by the rotation of the celestial sphere. Or, as we would put it, the earth was moving him from 0 to 0' by its rotation. In an hour his observatory moved eastward 15°, or about 500 miles ($\frac{1}{24}$ the distance around the earth at his latitude).

But the moon has its own eastward motion among the stars, as you have seen, completing a circle around the earth every $27\frac{1}{3}$ days, moving eastward among the stars about 13° per day or $5\frac{1}{2}$° each 10 hours. Way back in 300 B.C., Aristarchus had observed that it moved less than this and correctly attributed the lag to parallax. Tycho's very accurate measurments showed that the moon moved only $4\frac{1}{2}$° as he watched it during a 10-hour observing period. The 10-hour parallax of the moon was therefore one degree ($5\frac{1}{2}$° $-$ $4\frac{1}{2}$°).

The next evening when Tycho first observed the moon, it appeared in the sky the full 13° east of its position at the same hour on the previous night, showing that the lag was in the eye of the beholder — it was parallax. Tycho's comet was moving too; observations taken 24 hours apart showed it in different positions with respect to the stars. Comets show a parallax lag like that of the moon, except that the lag is much smaller. The exact angle for Tycho's comet was too tiny to measure. All he could be absolutely sure of was that the comet was at least three times farther than the moon. He believed that it was as far from the earth as Mercury and Venus, and that it crossed the orbit of Venus. After several weeks, the comet changed its motion among the stars and became fainter, but still had no detectable parallax lag.

From his daily observations of its angular distance east or west of the sun, Tycho reasoned that the comet's path came close to the sun. His data showed that the comet did not move at a uniform speed in this orbit. During his lifetime, Tycho also observed six other comets, and found that their orbits were, in his words, "not exquisitely circular, but somewhat oblong."

By his comet observations, Tycho demolished four revered principles of the Aristotle-Ptolemy camp, and the word got around:

(1) The planets could not be carried on crystalline spheres, as Aristotle thought, or even on solid rings and wheels, for then the comets could not move freely through the orbits of these planets.

(2) Not everything moves around the earth; comets move around the sun.

(3) Celestial motions are not all circular; comets move in elongated orbits.

(4) One type of celestial object, a comet, does not move with uniform speed in its orbit.

Nevertheless, Tycho did not accept the Copernican system. For him the chief stumbling block was the same bit of evidence that made Ptolemy sure that the earth isn't moving: no parallax among the stars. Each constellation appears equal and similar, not only from wherever you view it on earth but whenever you view it during the year. If the earth were moving in a yearly orbit around the sun, both

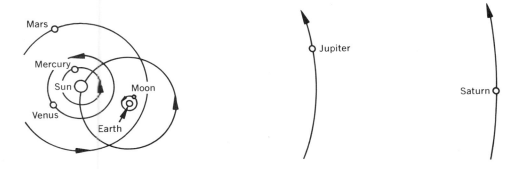

Fig. 5-4 Tycho's theory of the movement of the sun and planets.

men believed that parallax would change the outlines of the constellations, in the same way that a group of balloons on the ceiling appear to change their formation as you move around the room. Ptolemy, you will recall, argued that the identical appearance of the constellations from everywhere on earth indicated that "the earth is but a point in space," compared to the celestial sphere. It is odd that neither he nor Tycho went one step farther to conclude that the sphere of the stars is so big that even earth's movement in an orbit couldn't produce a parallax large enough for them to measure. Not only was the earth "but a point in space" compared to the celestial sphere, the earth's orbit was too.

Hoping to find stellar parallax, Tycho measured the positions of a thousand different stars over and over again with painstaking care. Because he found no parallax, he concluded that the earth must be fixed. But he compromised between the Ptolemaic and Copernican systems. He considered the earth at the center, the the sun revolving around it each year (Fig. 5-4), and the other planets traveling in circles around the sun in the same orbits that Copernicus gave them. The data used in Figure 4-6 to plot the orbit of Mars, according to Copernicus, is repeated in Figure 5-5. Notice that the relation of Mars to the starry background, as well as the angle between Mars and the sun, observed from the earth, is explained equally well by both theories. Tycho's circular orbits, of course, required some epicycles, as did those of Copernicus.

Tycho, The Observer

Copernicus' interpretation

Tycho's interpretation

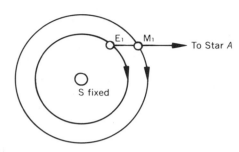
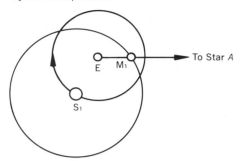

Feb. 15, 1967
Earth, Mars, Star A lined up;
Mars moving eastward among the stars.

Copernicus' interpretation

Tycho's interpretation

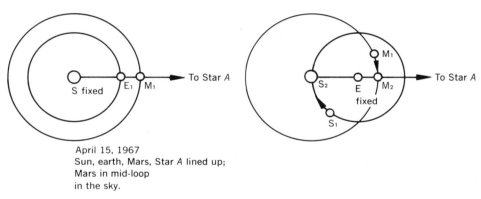

April 15, 1967
Sun, earth, Mars, Star A lined up;
Mars in mid-loop
in the sky.

Copernicus' interpretation

Tycho's interpretation

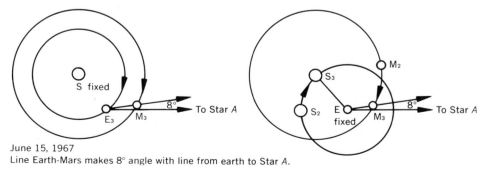

June 15, 1967
Line Earth-Mars makes 8° angle with line from earth to Star A.

Fig. 5-5 Copernicus' and Tycho's explanations of the movements of Mars in the sky, plotted for data of February 15, April 15, and June 15, 1967. (See also Figs. 4-5, 4-6.)

Tycho's theory is more complicated and more trouble to use. Because he had the earth as the central stationary body, however, his theory did not tread on as many toes as did the Copernican theory. People listened and became more willing to consider that Ptolemy's universe was not the only possible interpretation.

Additional Reading

DINGLE, HERBERT, "Tycho Brahe" in *Astronomy* (Samuel Rapport and Helen Wright, eds.): New York, New York University Press, 1964.
GADE, JOHN ALLYNE, *The Life and Times of Tycho Brahe,* Princeton, Princeton University Press, 1947.

chapter 6 | **One-Seventh of a Degree**

Tycho left Denmark in 1597, shortly after a new king ascended the throne. Tycho had been a despotic ruler over the subjects of his island and had ignored imperial court warnings to curb his abuse of power. Tycho's position was not helped by the fact that the king was under pressure to use Tycho's allowance for waging war with Sweden. When the new king trimmed Tycho's huge allowance to modest, but still adequate proportions, Tycho left Denmark in a huff and became Imperial Mathematician of Bohemia, at Prague, in what is now Czechoslovakia. There he spent the few remaining years of his life studying the data accumulated in his 20 years of observation.

In 1600, the year before Tycho's death, a young mathematician named Johannes Kepler became his assistant. Kepler, a former high school teacher in Austria, had become famous as an astrologer and prophet with his *Almanac* for 1595. He succeeded Tycho as Imperial Mathematician and spent the next 25 years altering the Copernican theory to agree with Tycho's observations, which were much more exact and far more closely spaced than those available to Copernicus. Tycho had begged his assistant to continue working with the newer theory that he had suggested (Fig. 5-4). Instead, Kepler wisely chose to work with the Copernican system. As we have seen (Fig. 5-5), the two theories explain motions in the sky equally well, but Copernicus' system is simpler than Tycho's. The task would be long and hard enough; no need to complicate it further.

Kepler had begun his work wondering why there were six planets, rather than five or seven or even twenty of them. One day while teaching, his interest was caught by a figure showing a circle that inscribed a triangle that in turn inscribed another circle. The two circles were in the ratio of the orbits of Saturn and Jupiter. He tried inscribing other polygons inside "Jupiter's orbit" to see if the circles inscribed therein would yield Mars' orbit. Perhaps, if he could find the right polygons, the inscribed circles would be in the ratio of all the planets' orbits. It didn't work.

Then it occurred to him that while you can construct an infinite number of regular polygons, there are only five regular solids (Fig. 6-1). Euclid had proved there could be no more. That explained why there were six planets. A great circle of the outermost sphere described the orbit of Saturn. A cube inscribed in this sphere circumscribed a sphere for Jupiter's orbit. A tetrahedron inside this held the sphere of Mars. A dodecahedron held the earth; an icosahedron, Venus; and finally, an octahedron, Mercury. It was beautiful. It reflected the perfect order of the Creator's mind. And it worked!

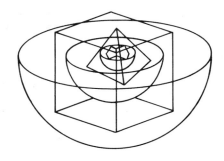

Fig. 6-1

Well, almost. There were many discrepancies, nearly all of them small. The fit wasn't quite perfect, but Kepler felt it would be if he had access to Tycho's observations. Tycho was pleased to have the young Kepler join him as an assistant. Tycho was growing old and he recognized in Kepler the genius that would be capable of completing his life's work. He assigned to Kepler the task of plotting the orbit of Mars, the planet whose orbit seemed trickiest.

As Kepler began his work, he found that Mars' position in the sky was 5° away from where the Copernican theory (now 60 years old) had predicted it would be. During the first years of his work, Kepler tried to fit various combinations of circular motions, including epicycles, to Tycho's observations of the positions of Mars. He was unsuccessful. Of course, they could be fitted if he piled on enough epicycles, just as an equation can be made to describe any series of points if enough terms are used. But if he had as many epicycles as Ptolemy's system, what advantage was there to the Copernican theory? At one point during these years of tedious work, he developed a modification of the Copernican system which had a reasonable number of epicycles. However, it disagreed with Tycho's observations by as much as eight minutes of arc (one-seventh of a degree). This was not good enough; he believed that Tycho's observations were not in error by as much as this, and so he discarded this solution to his problem.

Finally, perhaps remembering Tycho's comet orbits which were "not exquisitely circular," he gave up trying to fit all the observations of Mars to a circle. He simply decided to plot the positions of Mars and let them fall on whatever curve they might.

Selecting one of Tycho's observations of Mars (let us say for May 1, 1580), he plotted it as shown in Figure 6-2a. The angle between the lines of sight to the sun and to Mars, Tycho had observed to be angle x. The distance between the earth and the sun is E_1S; Mars is out in space somewhere along the arrow-tipped line.

Kepler knew that the period of Mars is 687 days (we called it about two years in Chapter 4). Therefore, Mars was back at the same place in its orbit in 687 days, on March 19, 1582. But where was the earth? Since May 1, 1580 it would have completed one revolution around the sun (365 days) and would be 322 days (687-365) along on a second revolution. It lacked 43 days (365-322) of being back at E_1. Since Kepler knew the angle through which the earth travels in 43 days (angle z), he was able to plot E_2, the position of the earth on March 19, 1582 (Fig. 6-2b). So he looked through Tycho's notes and found the angle in the sky between Mars and

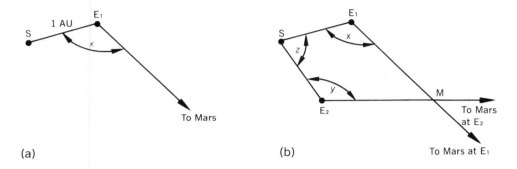

(a)   (b)

Fig. 6-2 Determining a point on the orbit of Mars by observations 687 days apart.

the sun on that day and plotted it (angle y). On that day Mars lay somewhere along the arrow-tipped line from E_2. But since he knew that Mars was at the same place in its orbit when the earth was at E_1 and at E_2, he knew that the point M, where the two arrow-tipped lines intersect, is where Mars was at both of these dates. M, then, is one position of Mars on its orbit.

He did the same thing for many other pairs of observations taken just 687 days apart. He then had a series of points, outlining the orbit of Mars. This curve is almost a circle, but not quite. Kepler found that the curve these points fit best is an *ellipse*.

The ellipse was felt to be a rather undistinguished figure, as compared to the perfection of the circle. More important, from Kepler's viewpoint, accepting elliptical orbits for the planets meant that he had to discard his beautiful theory of the five perfect solids. Yet he was able to do this rather than discard data, even a mere one-seventh of a degree.

The best way to define an ellipse is to describe how to draw one (Fig. 6-3). Attach the ends of a short piece of string to two tacks, and insert the tacks into a piece of paper on a drawing board, placing them fairly close together. Then push a pencil

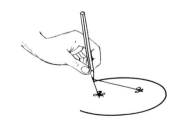

Fig. 6-3

against the string, keeping the string taut, and slide the pencil against the string, all the way around the tacks, back to where you started. The closed curve you draw is an ellipse. The longer the piece of string, the larger the ellipse. With the same piece of string, the farther apart you place the tacks, the more oval is the ellipse — the less the ellipse is like a circle. The closer the tacks are placed, the more circular is the ellipse, until, if the two tacks were in the same spot, you would draw a circle.

The ratio of the distance between the tacks to the string length tells how oval the ellipse is — how much it departs from being a true circle. This ratio is called the *eccentricity* of the ellipse. The larger the eccentricity, the more oval the ellipse. The eccentricity of all circles is 0. With a 10-inch string, tacks 2 inches apart give an ellipse of eccentricity $2/10$ (0.2); tacks 1 inch apart give an ellipse of eccentricity $1/10$ (0.1), about the eccentricity of Mars' orbit. When the tacks are 10 inches apart (eccentricity $10/10 = 1.0$), the string is tight between them, and the "oval" is as thin as it can get — a straight line between the tacks.

The location of each tack is called a focus. When plotting Mars in its orbit, Kepler found that the sun is at one focus. The other focus is empty. How surprising this must have been!

He also found that the orbit's curve did not form a perfect ellipse. Of course, this could be partly due to inaccurate timing of observations or measurements of angles in the sky. But Kepler recognized that there is another way in which larger errors could have been made. If Mars does not move in a circular orbit, he reasoned, it is possible that the earth doesn't either. In Figure 6-2b we have drawn SE_1 the same length as SE_2, as indeed they would be if the earth's orbit were circular. If the earth's orbit were elliptical, then SE_1 and SE_2 would be different lengths. This would shift their positions slightly, and this in turn would mean that the arrow-tipped lines to Mars would not cross at M, but at a slightly different place (Fig. 6-4). So Kepler tried ellipses of varying eccentricity for the earth's orbit and found that the observed positions of Mars fit well on his original ellipse if the earth's orbit is an ellipse of eccentricity almost 0.02.

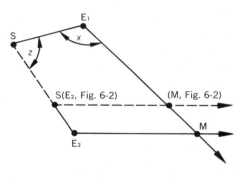

Fig. 6-4

This means that the "1," called the *astronomical unit* or *AU,* that we have taken for the distance of the earth from the sun is a sort of average distance. In one part of its orbit, the earth is slightly closer to the sun; in the rest it is farther. Copernicus had thought of Mar's orbit as a circle with radius 1.5 AU, but Kepler saw that this figure was Mars' average distance from the sun, or that the string length needed to draw Mars' ellipse must be 1.5 times the length for the earth's orbit.

In time, Kepler worked out the orbits for the other planets. All of Tycho's observations fit extremely well if the planet orbits are drawn as ellipses with the sun at one focus. The average distance of each planet from the sun is about equal to the radius that Copernicus found for his circular orbit (Table 2, col. 1). The eccentricities are small and different for each planet, varying from 0.006 for Venus to 0.2 for Mercury and 0.1 for Mars, as shown in column 2 of Table 2. You will see that the solar system, if drawn to a scale that would fit on the page, would appear to have circular orbits. It was fortunate that Kepler was working on Mars, whose orbit has a fairly large eccentricity. Had he been studying Venus he most certainly would have concluded that the orbits were circular.

The sun is not exactly at the center of these nearly circular orbits. At the focus, it is off the center by a fraction of the radius. As you can see by the way an ellipse is drawn, this fraction is equal to the eccentricity. So the sun is off center by a

	Approximate radius (AU)	×	Approximate eccentricity	=	Distance of sun from center of orbit (AU)
Mercury	0.4	×	0.2	=	0.08
Venus	0.7	×	0.006	=	0.004
Earth	1.0	×	0.02	=	0.02
Mars	1.5	×	0.1	=	0.15
Jupiter	5.2	×	0.05	=	0.25
Saturn	10.0	×	0.05	=	0.5

Table 2

different amount for each planet's orbit, as shown in column 3 of Table 2. These offsets are all in different directions and are quite large for the outer planets. Six eccentricities eliminated the 34 epicycles needed by Copernicus (and they never were needed again).

Kepler's orbits also eliminated the idea of constant speeds in perfect circles, and replaced it with a different "law" of planetary motion. Figure 6-5 shows the orbit of Mars; B, C, and D are located one-quarter, one-half, and three-quarters of the distance around the orbit. If Mars moved at uniform speed, the planet would go from A to B in about 172 days ($^{687}/4$). However, Kepler found that it goes beyond B in this time interval. On the other hand, Mars takes more than 172 days to go from B to C and also to travel from C to D. Then the planet goes from D to A in less than 172 days. Kepler found that Mars takes six days more to travel

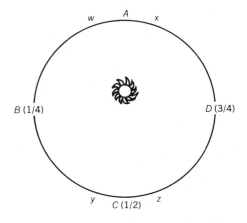

Fig. 6-5

the half orbit farther from the sun *(BCD)* than to travel the half orbit nearer the sun *(DAB)*.

More than this, he was able to show that each planet moves fastest in the part of its orbit between *w* and *x* and slowest between *y* and *z* (Fig. 6-5). He could pinpoint the speed change even more closely than this. Imagine a rubber band extending from the sun to the planet. As the rubber band shortens (as the planet comes closer to the sun), the speed of the planet increases. As the rubber band lengthens (as the planet gets farther from the sun), the speed decreases. As Kepler expressed it: The line between a planet and the sun sweeps over equal areas in equal lengths of time, as shown in Figure 6-6 (where the eccentricity is greatly exaggerated).

You can see that between March 21 and September 22 the earth moves halfway around its orbit. It completes the other half from September 22 to March 21. These are the two dates when the sun is seen at two opposite points in the sky, the two points where the ecliptic intersects the celestial equator (Fig. 2-4). There are three more days between March 21 and September 22 than be-

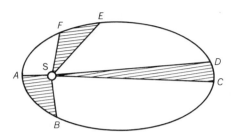

Fig. 6-6 A planet moves most rapidly when it is at *A*, nearest the focus of the ellipse where the sun is. The orbital speed varies in such a way that the planet moves through the shaded area *SAB* in the same time interval that it moves through *SCD*. The same is true of *SEF*. For this reason Kepler called this "the law of equal areas."

tween September 22 and March 21, showing that the earth goes a little slower in the first half and must, therefore, be farther from the sun.

In 1619, Kepler published a book called *The Harmony of the Worlds*. It is full of mystic nonsense of no lasting value. Kepler even went so far as to write down the notes of music played by the planets as they moved in their orbits, guided by angels. But it does contain another of Kepler's important discoveries, like a jewel tucked away in a box of excelsior. He was trying to find a connection between planetary orbits and the rules of musical harmony. Instead he found a surprisingly exact relationship between the planets' periods and their average distances from the sun (both based on Tycho's observations of positions in the sky). Kepler found that, for every planet, the square of the planet's period (in years) is equal to the cube of its average distance from the sun (in AU). This relationship is expressed by the equation $P^2 = R^3$ (Table 3). For instance, Jupiter's average distance from the sun is 5.2 AU. The cube of 5.2 is 140.6. According to the equation, the period of Jupiter should be the square root of 140.6, or just under 12 years, which agrees with the observations.

The Copernican system as revised by Kepler is an extremely neat and simple theory. More than that, the sun definitely appears to be something special, while the earth is just another planet. Although very few people realized it at the time, after Kepler the differences between Ptolemy's, Tycho's, and Copernicus' theo-

Planet	Distance (AU)	Period (years)	Distance cubed	Period squared
Mercury	0.387	0.241	0.058	0.058
Venus	0.723	0.615	0.378	0.378
Earth	1.00	1.00	1.00	1.00
Mars	1.524	1.881	3.54	3.54
Jupiter	5.20	11.86	141	141
Saturn	9.54	29.46	868	868

Table 3 Kepler's Harmonic Law.

ries were no longer merely those of geometry. As we have seen, all these systems could explain the motions of the sun and planets in the sky. Copernicus' was simplest to use, Ptolemy's offended people least, and Tycho's combined some of the advantages of both. But Kepler's work showed that there was a more fundamental difference between the Copernican theory and Ptolemy's. His studies indicated that motions of the planets were in some way dependent on the sun. He introduced an argument other than simplicity or taste for the correctness of the sun-centered universe.

Varying distance from the sun determines the varying orbital speed of each planet, including the earth. Although Copernicus (p. 32) had shown that the planets nearer the sun move faster in their orbits, Kepler showed that the harmonic law (Table 3) is exact. And you can see that the average speeds of planets in their orbits are not all the same. If they were, Jupiter at 5.2 AU from the sun would take only 5.2 years to go around the sun instead of 12 years. Because the orbits are almost circular, the distance around Jupiter's is about 5.2 times the distance around the earth's. If the planets moved at different average speeds, not dependent on their distances from the sun, but allotted in some haphazard manner (as indeed they were in the Ptolemaic rings and epicycles), no simple equation like $P^2 = R^3$ could give their periods in terms of their distances from the sun.

Of course, Tycho's theory could be defended by saying that the planets move around the sun in ellipses, while the sun, in turn, moves about the central earth in an ellipse. But then you have to explain the strange coincidence that the earth has exactly the same effect on the sun as the sun has on the five planets, in terms of varying speed along the orbit. And you must explain the further coincidence that the $P^2 = R^3$ law, which applies to all five planets going around the sun, also applies to the sun going around the earth.

Today we would consider that the sizes of the sun, earth, and planets might have something to do with which one dominates the others. It was known in 1600 that the sun is larger than the moon and farther from the earth, but all estimates of its size were much too low. In any case, few would have found it strange to think of a large body going around a smaller one.

47

Kepler's work accomplished two major steps that had not been taken in astronomy. The first was his acceptance of seemingly reliable data, even if the data did not agree with the hypothesis he was trying to confirm. The usual practice had been to discard such data as erroneous or irrelevant. And even Kepler did this on occasion.

Perhaps more important are the kinds of questions Kepler asked in the first place. Pre-Keplerian astronomers had been content to merely describe celestial motions, although they felt the accepted mechanisms were fictional. No one in Kepler's time actually believed in crystalline spheres and little wheels turning on big wheels. But Kepler asked why the planets' velocities should vary in different parts of their orbits and why the outer planets' velocities should be less than those of the inner planets. In assigning a physical cause to the planet's motion, he united astronomy and physics, and the two will never again be separated.

By asking the right questions and by following his data, Kepler discovered three laws that truly transformed astronomy from astrology:

(1) The planets travel in elliptical orbits with the sun at one focus.
(2) A ray from the sun to a planet sweeps equal areas in equal times.
(3) The square of a planet's year, or period, is proportional to the cube of its distance from the sun $(P^2 = R^3)$.

Kepler had to abandon the "perfection" of circular orbits and constant speeds. But in doing so he discovered a perfection and degree of order that had not been dreamed of. How important Kepler's laws were has perhaps been best stated by his biographer Arthur Koestler. He calls them the three "pillars on which Newton built the modern universe."

Additional Reading

> HART, IVOR B., "Johannes Kepler" in *Astronomy* (Samuel Rapport and Helen Wright, eds.): New York, New York University Press, 1964.
>
> KOESTLER, ARTHUR, *The Sleepwalkers* (Part 4: "The Watershed"): New York, Grosset & Dunlap (The Universal Library), 1963 or *The Watershed, A Biography of Johannes Kepler:* New York, Doubleday and Company, 1960.

chapter 7 | # Galileo's Telescope

In 1608, a Dutch spectacle maker put an eyeglass lens in each end of a tube. When he looked through the tube, distant objects appeared nearer and larger, showing details invisible to the eye alone. The telescope, as this device is called, was an interesting novelty, and many people saw that it could be useful in warfare and navigation.

In the early summer of 1609, the news of the invention reached the ears of an Italian scientist, Galileo Galilei, who immediately saw its importance to astronomy. He quickly bought two lenses from a local eyeglass maker and fitted them into a lead tube about 20 inches long. This telescope magnified objects about three times, about as effective as a pair of inexpensive field glasses. Amazed and delighted with what he saw as he turned it to the night sky, Galileo spent the next few months making a stronger instrument. By hand, he ground chunks of clear glass into lenses of the size and shape he wanted, and ended up with a telescope that magnified about 32 times. Using this instrument, he made several important discoveries.

The cross-section diagram in Figure 7-1 shows how a telescope makes a distant source of light appear brighter. Light rays passing from air to glass are bent, and they bend again as they leave the glass. The light from a star, so far away that its light rays are all parallel, enters the lens at one end of the telescope tube. Each ray is bent by a different amount because of the shape of the lens. If the lens is shaped just right, the bent rays all intersect at a point called the focus. They are all bent again by a second lens, the eyepiece, which acts like a magnifying glass for the observer's eye. The amount of light which gets into an eye unaided by the telescope is determined by the size of the lens in your eye. This lens is the pupil, about one-sixteenth of an inch in diameter. With the aid of a telescope, the light that falls on the whole telescope lens (Fig. 7-1) is focused to reach your eye.

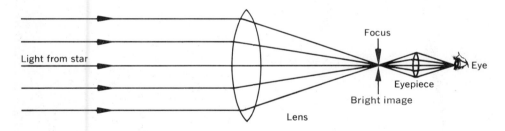

Fig. 7-1 How a telescope increases the brightness of a source of light.

Another way of looking at it is to say that in Figure 7-1 only one of the rays coming from the star will reach your eye, while the lens collects all of the rays drawn in the diagram and bends them in to reach your eye.

The telescope not only makes objects appear brighter, it also magnifies them as shown in Figure 7-2. Let us consider only the central rays coming from two stars. The angle from the stars observed in the sky by eye alone is $x°$. But notice that when observed through the telescope, they appear to be $y°$ apart. And y is a larger angle than x. If, instead of two stars, we had two points on opposite sides of a planet, the planet would appear larger. It is $x°$ wide as you saw it in the sky; it is $y°$ wide when viewed through the telescope. And the same is true for any two points on the planet's surface, or for any two points on, say, the surface of the moon.

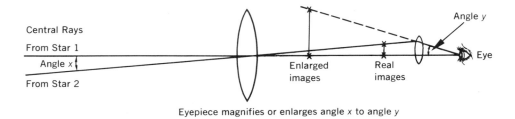

Galileo's Telescope

Eyepiece magnifies or enlarges angle *x* to angle *y*

Fig. 7-2 How a telescope magnifies a distant object.

Galileo was the first person to use a telescope to view the heavens. That is, he was the first person to leave a record of what he saw, to carry on a series of observations, and to realize the important evidence the telescope revealed about the universe. If, indeed, he were actually the first person to turn a telescope to the night sky, he had an experience almost unique in the history of mankind. It may be compared to that of the first astronaut, almost 350 years later. Standing on the earth, Galileo had a new view of outer space; traveling in outer space, the astronauts had a new view of the earth.

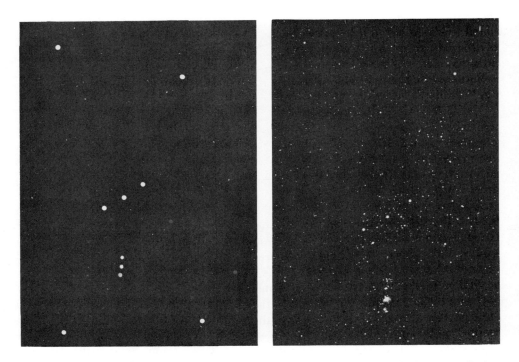

Fig. 7-3 (a) The constellation Orion, prominent in the evening sky in winter. (b) The constellation Orion, showing stars not visible to the unaided eye. [Yerkes Observatory photographs.]

The moon was the first object Galileo looked at with his instrument. Until that moment, most astronomers believed that the moon was a smooth globe, clear and perfect as a celestial object should be. The dark markings — the man in the moon — which everyone can see, were explained in many ingenious ways. One was that in these patches the moon's material was so "rare" that it let the light pass through, rather than reflecting it back. (Apparently they had never looked closely at a solar eclipse!) Galileo had realized that these markings must be on the moon's surface. But his telescope revealed more than just a mottled surface; he saw that the moon had great mountains and craters! Its surface is rough with large dark areas that he thought at first were seas. He saw that the moon resembles the earth; it is the same sort of object. Thus, the earth could be one of the celestial bodies. The Copernican system became more believable.

Next Galileo turned his telescope to the fixed stars and found that it did *not* make them larger. They were still points of light. However, it did show many more stars, stars too faint to be seen by eye alone. The telescope's lens brought enough more light to the eye to make them visible (Fig. 7-1). Compare the number of stars in Figure 7-3a, the naked-eye view of the constellation Orion, with the number of stars in Figure 7-3b, a photograph of the same constellation through a telescope.

Galileo's telescope also showed that some misty blurs of light in the sky are made up of many stars. One of these hazy patches, a star cluster called Praesepe, is in the zodiac constellation Cancer. You can find it in the sky with the help of Figure 2-6d. A good pair of field glasses will give you a view of it like Galileo had with his telescope.

Surely, while stargazing, you have noticed the Milky Way. It is a wide band of faint light extending diagonally across the sky, like a stream of skim milk. It forms a background to the constellations Cassiopeia, Cepheus, and the Summer Triangle (shown in Figures 2-6a through 2-6d), and extends all the way around the celestial sphere. About 300 years before Galileo's day, the poet Dante had said "...the Milky Way so gleameth white as to set the very sages questioning." Although the Milky Way has kept sages questioning up to the present, Galileo's telescope showed at least that the Milky Way is a multitude of faint, individual stars (Fig. 7-4).

All these stars never seen before showed the unreliability of ancient writers who did not even know that these stars existed, yet whose opinions about the universe were generally regarded as the final word.

When Galileo turned his telescope toward Jupiter, he saw that the planet was round like the sun and moon, and not a point of light like the fixed stars. He noticed three very small but bright "stars" near Jupiter, two to the east of the planet and one to the west of it. The odd thing was that these "stars" formed a straight line with Jupiter, and this line was parallel to the ecliptic. The following night he was surprised to see that all three of the "stars" were west of Jupiter. The next night he saw only two of the "stars." He looked at them through his telescope for many nights, sketching their positions in his notebook. A portion of his record is shown in Figure 7-5. You will notice that sometimes he saw four

Fig. 7-4 Telescopic photograph of the Milky Way in the area of the constellation Cygnus. [Yerkes Observatory photograph.]

"stars," sometimes three, and sometimes only two. He noticed that they were always near Jupiter, as that planet moved against the background of the fixed stars. These observations showed Galileo that the four "stars" revolve in orbits around the planet Jupiter. They are Jupiter's moons, or *satellites*. As we watch each satellite circle Jupiter, it is sometimes behind the planet and we can't see it. Sometimes it is at one side of Jupiter, sometimes on the other side. And sometimes it is in front of the planet, invisible against the bright disk. Galileo's observations showed that the moons move with great regularity. He determined that the period of each one is different, varying from just under two days to about 17 days.

Jupiter, with its moons, is a small-scale model of the Copernican system. It looks like the solar system would if we could view it from outside. And there is nary an epicycle in evidence! Galileo's discovery answered a common objection to the Copernican theory, that our moon is a strange exception because it alone moves around a planet instead of around the sun. It also proved, more clearly than Tycho's comets, the error of the ancient doctrine that the earth is the only center of motion in the universe. It laid to rest the argument that if the earth were moving, the moon would be left behind because it could not be expected to keep up with the earth, rapidly orbiting the sun. Jupiter's satellites were having no difficulty staying with Jupiter!

Galileo was the first to see that Venus has phases like our moon does. As he observed that planet night after night, he saw that Venus went through crescent, half, gibbous, and full phases. This showed that Venus does not shine by its own light, but by reflecting sunlight. He believed that this must also, then, be true of the other planets. On hearing this news, Kepler wrote Galileo that it now looked as if the fixed stars were suns and the planets were, as he expressed it, "earths."

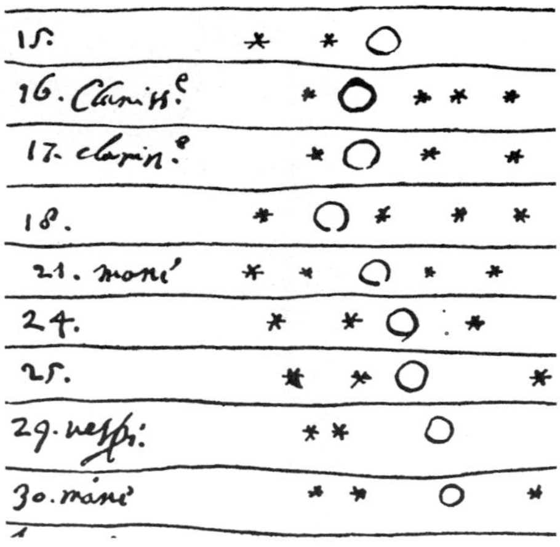

The phases show that Venus is moving all the way around the sun, from the near side to the far side, when it looks like a

Fig. 7-5 Galileo's drawings of the moons of Jupiter. [From *The Starry Messenger,* courtesy of Yerkes Observatory.]

full moon, only much smaller. Remember that Venus is never seen more than 47° from the sun in the sky, so it can't be opposite the sun (180° from it) like the full moon is. The full phase of Venus, seen when it is close to the sun in the sky, shows that it is then beyond the sun; the thinner and larger crescents show that Venus comes around between us and the sun the next time we see it close to the sun in the sky. This confirms the orbit as Copernicus (and Tycho Brahe) assumed it. Figure 4-4 shows similar phases of Mercury, explained in the same way. In neither case can the planet always be on our side of the sun, in orbit around the earth, as Ptolemy thought.

With a good pair of field glasses or a small telescope, you can see what Galileo saw. But never look through field glasses or a telescope at the next object that Galileo examined. He looked at the sun with his telescope, a very dangerous thing to do, for it often causes permanent damage to the eyes. (In fact, Galileo went blind later on.)

What he saw is shown in the modern photograph in Figure 7-6: The sun has blemishes. These *sunspots,* as they are called, are now known to be large, cooler areas on the sun that appear dark in contrast to the brighter and hotter solar surface around them. Sunspots are temporary, usually lasting only a few weeks to a few months. Large sunspots had been observed before Galileo's time with the eye alone, but were generally regarded either as something in the earth's atmosphere or somewhat nearer the sun, silhouetted against its bright surface. However Galileo watched the sunspots for many months and saw that they moved in a regular manner across the disk, to its edge, and then disappeared. He concluded from this that the sun must be rotating, carrying the current sunspots into our view and then out of sight again. Sunspots were a striking exception to Aristotle's

53

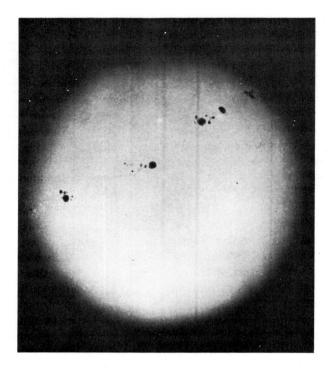

Fig. 7-6 Photograph of the sun, taken on October 13, 1926 by W. W. Morgan with a telescope having a 12-inch lens. [Yerkes Observatory photograph.]

belief in the perfection of celestial objects. Moreover, the rotation of the sun made earth's rotation seem more possible.

Each of Galileo's discoveries was a blow struck at the Ptolemaic system and at the ideas of Aristotle. But if Aristotle and Ptolemy had been standing at his elbow as he made these discoveries, they would probably have been eager to look through his telescope — unlike their seventeenth-century followers, many of whom refused to. Aristotle and Ptolemy probably would have been the first to admit that they were wrong. The system of the universe that Ptolemy and the earlier Greek philosophers had devised was based on their observations and it explained these observations adequately. The scholars who taught the Ptolemaic system as absolute truth in 1600, however, accepted it blindly. They looked on Galileo as a troublemaker, out to upset the established order and to question their authority. We look on him today as one of the founders of modern science, who reintroduced experiment and observation as the basis for scientific inquiry rather than hashing over the opinions of authorities, with never a look at nature. Kepler as we have seen, also considered observational data as the final authority.

In 1610, Galileo published a book, *The Starry Messenger,* which described his observations with the telescope. The book was widely read but was most unpopular with the church authorities. Primarily because of it, the Pope declared the Copernican doctrine "false and absurd," and anyone who taught that it was true was to be punished. Nevertheless, Galileo followed his first book with a second, *Dialogues Concerning the Two Principal Systems of the World,* a crushing attack on Ptolemaic astronomy and a convincing argument for the Copernican system. He was so sure of his arguments that he wrote a long letter to one of the Cardinals in Rome, saying that any other opinion was stupid. The result was that Galileo was brought before the Inquisition in 1632 and threatened with torture although he was 70 years old at the time. He did formally "abjure, curse, and detest" his "errors and heresies," but as he left the court he is reported to have said (in Italian), "...but the earth does move!"

Part of the reason for his recantation was that his work was not yet finished. He spent the eight remaining years of his life as a prisoner in his home, secretly writing his most admired and perhaps most valuable work, *Two New Sciences,* about motion and the structure of matter. He felt, rightly, that the truth or falsity of the Copernican system was no business of these "authorities" who had never looked through a telescope, never observed a planet's motion across the sky, never drawn a planet's orbit, and never considered the evidence with open minds. He made his point. Scientific truth cannot be decided by popular vote or by the opinion of any authority. It can be decided only by observation or experiment, and by reasoning sensibly about what is seen. Two hundred years later, early geologists in a milder conflict with the authorities of established religion put it this way: "We shall let earth herself tell us her story." And, armed with rock hammers, magnifying glasses, and a theory, they went to work to decipher the history and age of our planet, the earth.

Additional Reading

GALILEI, GALILEO, "Telescopic Observations" in *Astronomy* (Samuel Rapport and Helen Wright, eds.): New York, New York University Press, 1964.

KOESTLER, ARTHUR, *The Sleepwalkers:* New York, Grosset and Dunlap (Universal Library), pp. 352-496.

PAUL, H. E., *Telescopes for Skygazing:* Cambridge, Mass., Sky Publishing Corp., 1960.

chapter 8 | Newton's Explanation

Galileo's observations and Kepler's calculations did make it seem likely that the Copernican theory is the true picture of the solar system. But a serious question remained unanswered. What had kept the earth moving around the sun during the long time that men had been observing the sky? When it was thought that the sun, the planets, and the sphere of the fixed stars were in motion, rather than the earth, there was no problem. It was generally believed that because they were celestial objects, they could move forever without effort.

The Greek philosophers had also taught that, on earth, rest is the natural state and that it took effort, a *force,* to put things in motion and keep them moving. Instead of merely accepting this opinion, Galileo carried out experiments with moving objects. He found that when a wooden block is slid across the floor, it soon comes to rest. The friction with the floor acts as a retarding force. But when he reduced the friction by using a polished block on a very smooth floor the block slid farther before coming to rest. (On smooth ice, this polished block slides still farther, because there is even less friction.) Galileo's experiments showed that the smaller the retarding force is, the less the block tends to slow down, and the farther it moves before coming to rest.

Although he couldn't prove it he reasoned that if *all* friction could be removed — if no resisting forces acted on the block — it would continue in motion at the same speed forever. His experiments showed him that the Greeks were only half right. A force is required to *start* the motion, but not to keep the block moving. In fact, he found that a force is necessary in order to slow it down, speed it up, or change the direction in which it is moving. But a force is *not* necessary to keep an object moving in a straight line at a constant speed. Galileo's experiments convinced him that rest is no more natural than is straight-line motion at uniform speed. He called this the *principle of inertia.*

Galileo believed that there is no fundamental difference between celestial objects and earthly ones. (Recall that he had discovered that the moon is not a perfectly smooth sphere.) Therefore, his principle of inertia could apply to the movements of the planets as well as to blocks sliding along a smooth floor. Once started in motion, the earth and the other planets would keep moving. This would explain their continued motion for thousands of years. However, as you are probably already saying, it certainly doesn't explain why they move in ellipses at changing speeds, as Kepler found. Why don't they move "naturally," at constant speeds in straight lines? Why do they keep going around the sun instead of moving along straight paths that would long ago have taken them far away? If Galileo's principle is right, and if it applies to the planets, then a force must be continually acting on them. What is this force? Galileo did not deal with this question, perhaps because there was not enough time before his death to do so.

However, he did answer an old objection to the Copernican theory. People had long said that if the earth were moving, objects on its surface would be left behind. Birds flying through the air and the air itself, couldn't keep up with the moving earth. Galileo pointed out that if you drop a stone from the mast of a ship lying at anchor, the stone falls to the deck at the base of the mast. If you drop the stone from the mast while the ship is moving, it still falls to the deck at the base of the mast. This shows that the stone shares the forward motion of the ship. If the stone did not share this motion, it would land in the sea behind the moving ship, or at least on the deck back of the mast. In a similar way, objects on the earth, birds in the air, and the air itself share the earth's motion, and are not swept off or left behind as the earth moves, any more than the stone is left behind the ship. A force has to act on the stone (you have to throw it, rather than merely drop it) before it falls anywhere but at the base of the mast. This is because it takes a force to change the forward motion of the stone, as the principle of inertia predicts.

The stone fell until it landed on something that could support it, in this case, the deck. If you stand in a field and throw a stone straight up, it always falls back to earth. Moving upward, it gradually slows down, then stops, and then starts speeding up again as it moves back toward the earth. Galileo watched things fall (some say that he dropped objects from the top of the Leaning Tower of Pisa), and found that freely falling bodies are *accelerated*; they gain speed as they fall. Moreover, he observed that all of them speed up by the same amount in equal intervals of time; the acceleration is always the same. This increase in speed amounts to 32 feet per second in each second. A stone dropped from a tower is moving at a speed of 32 feet per second at the end of the first second, at a speed of 64 feet per second at the end of the second second, at a speed of 96 feet per second at the end of the third second, and so on, until it lands on the ground.

The most surprising thing Galileo found is that whatever the weight or size of the object, it falls with the same acceleration. At the end of each second, all falling objects move at the same rate. It had always been taught that heavier objects fall faster than lighter ones, but Galileo was the first person to drop an assortment of objects, time their fall, and find out. (Of course, light thin objects like a feather or a leaf fall slower than more compact objects because of air resistance, a force opposing their downward motion. They can also be blown around by the wind. This force of the air is used by a parachute jumper. The air resistance on a big parachute pulled at 32 feet per second is about 200 pounds.)

Galileo's data on falling bodies and his principle of inertia (as well as Kepler's elliptical orbits, his law of areas, and his harmonic law) were all known to Sir Isaac Newton. Newton was born in England in 1643, the year after Galileo died. He graduated from Cambridge University at the age of 22 and went back to his home, where he spent the next two years studying the question of why the planets move eternally in elliptical orbits around the sun. One of the anecdotes that almost everyone knows tells how he started on this problem as he watched an apple fall. Unlike the tale of George Washington and the cherry tree, this appears to be a true story.

Newton asked himself why the apple fell downward. Why didn't it go sideways or upward? It must be, he thought, that the earth attracts it. Galileo had shown that the apple, or any other falling body, is accelerated — its downward speed steadily increases. The principle of inertia says that if there is no force on such a body, it moves at constant speed. Newton saw, therefore, that a force was acting on the apple. Because the apple fell toward the earth, it seemed likely that this force had something to do with the earth. It was an odd kind of force. Unlike a push on the apple given by hand when you throw it, this force toward the earth acted without any connection. No hand or baseball bat or rope is needed to force the apple downward as it falls. Such a force could work on other bodies far from the earth, such as the moon. Of course, today we know that it works on airplanes, missiles, and artificial satellites. Newton called the force *gravity*, and, in a flash of genius, decided that it was the material of the earth pulling on the material of the apple.

Since both the earth and the apple are made of matter, it seemed reasonable that the size of the force must depend on the mass, or quantity of matter, in both of them. Newton reasoned that the force of gravity must be proportional to the product of the two masses, or $F \propto mM$, where m is the mass of the falling object and M is the mass of the earth.

You can see that if you use a larger apple so that m is doubled, the gravitational force toward the earth is twice as large. At first glance, it looks as though the acceleration of this larger apple would also be doubled. Yet this is not so. Galileo found that all freely falling bodies have the same acceleration. This is taken care of by Galileo's principle of inertia, which Newton wrote as $F = ma$. F is any force acting on mass m, and a is the acceleration of m that it produces. If the mass of the apple is doubled, then the force of gravity on it is doubled. But when both the force and the mass are doubled, Newton's equation $F = ma$ shows that the acceleration remains the same. That is, when ma is substituted for F in $F \propto mM$ (that is, $ma \propto Mm$), the mass of the apple cancels out and $a \propto M$. So far, then, these two equations explained Galileo's observation that all bodies fall with the same acceleration.

Why did Newton consider the force of gravity as proportional to the product of the two masses ($m \times M$) rather than, for instance, as proportional to their sum ($m + M$)? Let us consider the apple twice as big. Because the mass of even this larger apple is extremely small in comparison to the earth's mass, the sum $m_2 + M$ is almost exactly the same as $m_1 + M$ for the first apple. So the predicted gravitational forces on the two apples would be almost the same if we merely added masses. In the equation $F = ma$, F would be only a very tiny bit larger, while m_2 is twice as great as m_1. Therefore, one would predict the larger apple's acceleration to be just about half that of the smaller one's, and this is contrary to what Galileo found out about falling bodies. Only if the force of gravity is proportional to the product of the interacting masses can all falling bodies near the earth's surface be speeded up by the same amount in equal time intervals.

Because the force of gravity between two masses depends on both of them there is no reason to suppose that it would only accelerate the smaller object. Newton reasoned that as the earth pulls the apple down toward it, the apple pulls the earth up toward it. The force acting on each is equal, but in opposite directions. The acceleration of the very massive earth is too small to be detectable, just as the equation $F = ma$ predicts.

Newton, like Galileo, believed that the same natural laws that apply on earth apply also to the planets, sun, moon, and stars. The movement of the planets in their elliptical orbits showed him that a force is acting on them. Because they are material objects, a gravitational attraction is to be expected between the sun and each planet. Yet the planets do not seem to be falling into the sun, like the apple falls to the earth. Newton recalled the motion of a thrown stone, like that in Figure 8-1. As it moves forward, the stone is constantly falling toward the earth. The

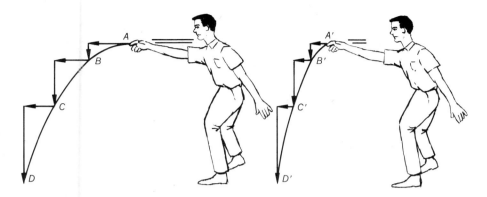

Fig. 8-1 The lightweight, broken lines show the paths of two stones thrown sideways at the same moment. The man on the left threw his stone harder than the man on the right and, therefore, the stone on the left has the greater forward speed. The horizontal arrows at A and A' show where each stone would have arrived after the same interval of time t, if gravity had no effect. However, the force of gravity pulled them downward by the amount shown by the vertical arrow. So, at time t, the stones are at B and B'. Both have fallen through the same vertical distance. At 2t, the end of another interval of time equal to t, they are at C and C', for gravity has pulled them downward by three times the amount as it did in time t. At time 3t they hit the ground at D and D'.

harder the stone is thrown, the farther it travels before falling to the ground. Newton could explain this curved motion in terms of gravity continually acting on the stone, as it moved forward with the speed imparted to it by the person who threw it.

If the stone could be given a high enough forward speed, he reasoned, it would completely encircle the earth, always falling but never reaching the earth's surface. Its curved path around the curved earth would keep them the same distance apart. If the moon had somehow been given a large sideways speed, then earth's gravity would keep it moving in its orbit around the earth. If the planets were given motions fast enough at right angles to the sun, the sun's gravitational attraction would keep them moving in their orbits, always falling toward the sun.

Figure 8-2 shows the sun and a planet. At point A, the planet is moving in the direction of the larger arrow. The force of gravity between the planet and the sun

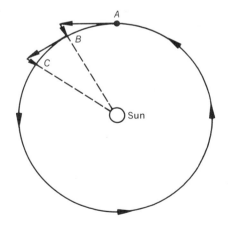

Fig. 8-2

59

is pulling the planet toward the sun, as shown by the shorter arrow. The result is that the planet moves to point *B*. At *B* the planet tends to keep on moving in the new direction that took it to *B*. But the sun's gravity brings it down to *C*, and so on, around the orbit. In Figure 8-2, points *A*, *B*, and *C* should be drawn very close together (as indeed they should also be in Figure 8-1). At every instant, the force of gravity is accelerating the planet. Therefore, its direction of motion is constantly changing. At every instant, the gravitational attraction between the planets and the sun provides the proper force to accelerate the planet in its elliptical path.

Kepler had found that the planets move fastest along their orbits when they are nearest the sun. This means that they are changing direction and speed most rapidly there—the acceleration is greatest. He saw that the planets in orbits close to the sun, like Mercury and Venus, are accelerated more than those farther from the sun. It was clear to Newton, therefore, that the force of gravity between the sun and a planet was somehow dependent on the distance of the planet from the sun. The farther apart the two masses are, the less is the gravitational attraction between them. From Kepler's law of areas and his harmonic law, Newton calculated that the force of gravity between two masses is inversely proportional to the *square* of their distance apart, or that $F \propto 1/R^2$ where F is the force produced by gravitational attraction and R is the distance between the two masses. In other words, the masses accelerate each other four times as much if the distance between them is cut in half.

An analogy will show that this is reasonable. The farther you are from a source of light, like a candle flame or a streetlight, the less light you receive from it. As the light rays spread out in all directions, we can think of them as spread over the entire surface of each of a series of spheres. The area of each sphere is $4\pi r^2$, where r is the sphere's radius. If you move twice as far from the candle, you are moving onto a sphere with twice the radius. The light is spread over an area four times greater. The light received at any point on that sphere is one-quarter as bright. The intensity of the light at any point on any sphere is inversely proportional to the square of its distance from the candle; it is proportional to $1/r^2$. In the same way, the intensity of the force of gravity is inversely proportional to the square of the distance between two masses. You can guess that this inverse-square relationship is simply the result of the geometrical formula for the area of a sphere.

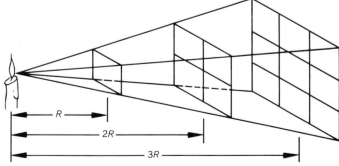

Fig. 8-3 The brightness of a source of light becomes four times less as you move twice as far away; nine times less as you move three times away.

From this inverse-square relationship you would predict that the force on the apple increased as the apple fell closer to the earth. Because the apple fell through such a short distance compared with the larger size of the earth, the change in its acceleration is too small to be measured. But Newton found that the moon is accelerating only $1/3600$ times as fast. The moon is 60 times farther from the center of the earth, so its acceleration toward the earth is $1/60^2$ as much as the apple's.

Newton combined $F \propto mM$ with $F \propto 1/R^2$ into his *law of universal gravitation*, $F \propto Mm/R^2$. He said that every particle of matter in the universe attracts every other particle with a force F which is directly proportional to the product of the masses of the two particles mM and inversely proportional to the square of their distance R apart.

Newton said that this law applies to "every particle of matter in the universe." Of course, he couldn't prove it. The law says that the kitchen stove and the kitchen table attract each other. They don't seem to. He was confident that this is because the gravitational forces between two such small masses are so very tiny that they are swamped by other much larger forces, such as the earth's gravity. His confidence was justified. About 130 years later, careful laboratory measurements did reveal the tiny gravitational attraction between two small lead spheres. The experiment was performed by another Englishman, Henry Cavendish.

Newton could prove neither that his law of gravitation nor his law of inertia $(F = ma)$ apply to the whole universe—to the fixed stars and whatever might lie beyond them. But all the bodies in the solar system moved as these laws predicted, and he saw no reason to believe that other celestial objects are different.

Newton's simple laws explain a wide variety of motions, from the falling apple to the moon in its orbit and the planets in theirs. Together with Galileo's telescope, they enabled astronomers to explore the universe in new ways. As we shall see, they revealed a universe whose size, variety, and splendor was undreamed of.

Additional Reading

ASHFORD, T. A., *From Atoms to Stars:* New York, Holt, Rinehart and Winston, 1960, pp. 50-69.

BELL, E. T., "Isaac Newton," in *Astronomy* (Samuel Rapport and Helen Wright, eds.): New York, New York University Press, 1964.

chapter 9 | **"But the Earth Does Move!"**

In about 150 years, five men had erased the picture of the universe that had been accepted almost without question for more than 14 centuries. Copernicus' idea, Tycho's accurate measurements, Kepler's persistence, the imagination and clear thinking of Galileo, and Newton's brilliant theory had replaced the time-honored

"But the Earth Does Move!"

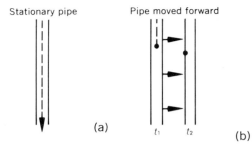

Stationary pipe   Pipe moved forward   Pipe moved forward at proper tilt

(a)   (b)   (c)

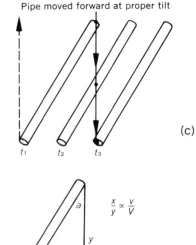

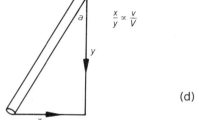

$$\frac{x}{y} \propto \frac{v}{V}$$

(d)

Fig. 9-1 (a) A raindrop entering a motionless vertical pipe will fall straight through the pipe. (b) A raindrop will not fall through the central opening of a vertical pipe which is being moved. Between times t_1 and t_2, the pipe moved forward as shown and the raindrop moved downward. Consequently, the raindrop collided with the side of the pipe. (c) Now if the pipe is tilted from the vertical through angle a as shown, the raindrop falls through the central opening of the pipe. (d) A raindrop will fall all the way through the pipe if $x/y = v/V$, where v is the velocity with which the pipe is being carried forward and V is the downward speed of the raindrop.

picture with a new one. The idea that the earth revolves around the sun, considered ridiculous in the sixteenth century and immoral in the seventeenth century, came to be taken as a matter of course in the eighteenth century.

In the nineteenth century, parallax of the stars, sought for since Ptolemy's day as proof that the earth is moving, was at last observed. But it was almost an anticlimax. By then, astronomers had made many new discoveries about the universe, all of them consistent with Newton's laws, some of them even predicted by those laws. And in 1727, an English astronomer named James Bradley presented a proof of the earth's motion that no one had even thought of before.

On a calm, rainy day, as you wait for a traffic light to change, raindrops fall straight down past the car window. But when the car starts moving again, the drops seem to be falling somewhat diagonally past the window, as if they were coming from a slightly forward direction. When you stop the car at the next light, however, the rain is falling vertically again. The raindrops were falling straight down all the while, but an effect called *aberration* made them appear to slant toward the rear of the moving car.

Bradley may have noticed the same effect from a stagecoach window. At any rate, aberration of raindrops suggested to him a way of proving whether or not the earth is moving. An analogy, illustrated in Figure 9-1, will show his line of argument. If you stand still, holding a length of hollow pipe vertically, a raindrop which enters the pipe will fall straight through it (a). But if you start walking, the raindrop can't fall straight through the pipe; it hits the inside of the pipe before it reaches the bottom, as shown in (b). However, if you tilt the moving pipe just right, as in (c), the raindrop falls all the way through the pipe. The tilt is right (d) when the distance *x* that the top of the pipe precedes the base of the pipe, divided by the vertical distance *y* between the top and bottom of the pipe, is equal to the speed at which you are walking *v*, divided by the downward speed of the raindrops *V*. More simply put, a raindrop will fall through a pipe tilted so that $x/y = v/V$.

If you are walking in the opposite direction, the tilt must be reversed. If you are walking faster, it must be increased. If the rain is really slanting toward you, as in a windstorm, the pipe does not have to be tilted as far away from the direction in which the rain is coming. If the rain were blowing along parallel to the ground, you could point the pipe straight at it and catch the raindrops.

A telescope catching light from a star can be compared to a pipe catching raindrops. If the earth isn't moving, you can point the telescope directly toward any star and its light will move down the telescope tube and reach your eye. But if the earth is moving, you must tilt your telescope away from the direction in which the light is coming, or the light won't travel all the way down the tube.

The earth's movement in an almost circular orbit, Bradley saw, would mean that the direction in which the telescope is tilted from a star's true direction would change continually throughout the year. And because there are stars in all directions, the tilt wouldn't be the same for all of them at any one time. Neither the *direction* nor the *size* of the angle of tilt would be the same for all stars, and this is in addition to any parallax. Light from stars 90° north of the ecliptic, for instance, comes straight down on the earth's orbit, like the raindrops in Figure 9-1. For these stars (Fig. 9-2), the tilt away from the true direction would be largest. Stars "dead ahead," toward which the earth is moving on a given night, would not be shifted at all. The light from other stars comes in at various angles to the direction of the earth's motion, and the amount of tilt necessary for each one would be different, but predictable.

If the earth is moving in an orbit, all the stars would appear to be continually shifting position on the celestial sphere, each one appearing to move by a slightly

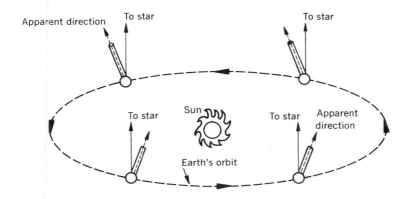

Fig. 9-2 Aberration of the light from a star 90° above the plane of the earth's orbit.

different amount in a slightly different direction. In order to see whether they did or not, Bradley had to measure angles between many stars on many nights. He had to measure precisely, for the changes were very small. In the end, he found that all stars do appear to move, relative to each other, in the amounts and directions predicted for aberration. The largest shift in position of any star — the largest tilt necessary — is $20\frac{1}{2}$ seconds of arc (20.5"), about $\frac{1}{180}$ of a degree.

In Figure 9-1d if the angle a is 20.5", the ratio of x to y is 1 to 10,000, and v/V is in the same ratio. This means that the speed of light V is 10,000 times that of the earth v. Fifty years before, a Danish astronomer named Ole Roemer had measured their relative speeds, using the periods of Jupiter's moons as a sort of clock, and found this same ratio. Thus, all the evidence fitted together, and the orbital motion of the earth appeared to be a reality.

As the Copernican idea came to be generally accepted, the fact that no star showed parallax began to be considered as evidence for the tremendous distances of the stars. As we have seen (Fig. 5-3), the nearer an object is, the larger is its parallax. If the stars were all at the same distance, the parallax of each would be the same amount, but the positions of the stars would appear to change in different ways, according to their position on the celestial sphere. Thus as the earth travels in its orbit around the sun, a star lying in the plane of the earth's orbit appears to shift back and forth in a straight line as the earth passes from one side of the sun to the other. A star lying directly above the plane of our nearly circular orbit, however, appears to move in a small circle once each year. Its yearly parallax is the diameter of the circle. A star between these two extremes appears to travel yearly along a small ellipse; the long dimension of the ellipse is its yearly parallax. Of course, the effect of aberration must be allowed for in searching for parallax.

This search continued, and in 1838 three astronomers, working independently, were able to measure the parallaxes of three different stars. Thomas Henderson, at the Cape of Good Hope in Africa, found that the yearly change of position of a star called Alpha Centauri is 1.5 seconds of arc (1.5"), only $\frac{1}{2400}$ of a degree. It is no wonder that Tycho couldn't find stellar parallax without a telescope, for Alpha Centauri has just about the largest parallax of any star. It comes very close to being the star nearest the earth. Because it is only 30° from the south celestial pole, it is not visible from the United States or from most of Europe.

The other two men measured the yearly parallax of different stars. F. W. Bessel, in Germany, found the parallax of a star (visible only through a telescope) in the constellation Cygnus (Fig. 2-6a) to be only 0.6"; Wilhelm Struve, in Russia, measured that of Vega to be 0.25".

Since then, parallax measurements have been made on about 700 stars. But these first measurements (or, indeed, any one of them) prove that the earth orbits the sun. The apparent movements discussed earlier repeat themselves annually; and the greatest differences in position are between observations made six months apart, since it takes the earth six months to move to the opposite side of its orbit. The different amounts of parallax of these three stars show that they are at different distances from the sun. Vega, with a parallax six times smaller than that of

Alpha Centauri, is six times farther away. Notice that this very small parallax was measured by comparing the position of a star with positions of fainter background stars assumed to be much farther away, so that their parallaxes are about zero.

The fact that parallax differs so widely shows that the stars are scattered through space, rather than being on a celestial sphere all at the same distance. As we look out into space, the sky appears to be a half-sphere and the horizon to be a circle. This is because we see equally well in all directions. The stars only appear to lie on a sphere; their positions in the sky are not definite locations, but merely directions, or lines of sight to each star. The flashing lights of an airplane can have the same position on the celestial sphere as a far-off planet or an even more distant star, if all three are in the same direction from us. Although astronomers describe the direction of an object in the sky in terms of its location on the celestial sphere, after 1838 the sphere could no longer be considered a reality. And a sphere that isn't there can't rotate. The observed daily motions in the sky must be due to the earth's daily rotation. The measurements of stellar parallax certainly suggest that the earth is rotating, as well as revolving yearly about the sun.

Galileo had shown that the earth *could* be rotating—the fact that things are not flying off its surface or being left behind does not prove that the earth is not rotating; but neither does it prove that it is. There was no direct evidence that the earth rotates until 1851, when a French physicist named Jean Foucault fastened one end of a long wire high in the dome of a building in Paris and attached a heavy metal ball to the other end. This ball and wire formed a very long pendulum, somewhat like the much smaller pendulum of a grandfather clock. The base of the metal ball had a sharp point, and Foucault had placed a ring of loose sand on the surface beneath it. He started the pendulum and let it swing back and forth. It crossed and recrossed the ring of sand, and at each crossing made a mark in the sand.

Once he had set the pendulum in motion, the only force acting on it, except for a very small air resistance, was gravity. And this force was in a downward, not sideward, direction. Air resistance would slow it up a little, but there was no force acting on the pendulum which could change its dirction of swing. So according to Newton's law, $F = ma$, the pendulum should continue to swing in the same direction.

Yet it soon became apparent that the swing of Foucault's pendulum was shifting gradually to a new compass direction. He saw two broad bands of marks opposite each other on the ring of sand. At first the line of swing was AA' (Fig. 9-3); gradually it moved to BB', after a longer time interval to CC', and so on. The plane of swing had shifted with respect to the parts of the room, to landmarks outside, or to compass directions.

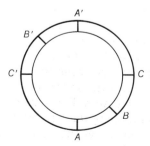

Fig. 9-3

"But the Earth Does Move!"

Foucault concluded that the pendulum only appears to change its direction of swing. It appears to do so because the earth is turning underneath it. Imagine swinging a chain and locket (the pendulum) back and forth over a phonograph record (the earth) as it plays. The swing of the chain keeps the same direction with respect to the furniture in the room (the fixed stars). However, an ant on the record would see the line of swing changing direction continually with respect to marks on the label underfoot (a landmark on the earth). It is the record that is moving, not the room and not the line of swing of the locket.

The line of swing of Foucault's pendulum in Paris appeared to rotate through 11° each hour. Therefore, it did not stay in the same direction with respect to the stars; each star appears to rotate around the earth once each day, or at the rate of 15° per hour. When pendulums like his were set up in other cities, it was seen that the apparent rotation of their lines of swing lagged behind that of the stars by different amounts. Those installed farther north lagged less; those farther south, more. Further study showed that the hourly rate increased with increasing distance from the equator, as measured along a line parallel to the earth's axis (Fig. 9-4a). When these distances are plotted against the observed hourly rates of pendulum rotation, as in Figure 9-4b, the points all fall on a straight line.

Even though no pendulums like Foucault's have been installed north of Uppsala, Sweden, or south of Singapore, we can predict the hourly rotation rates of a pendulum's line of swing at the North Pole and at the equator. Because there is no reason to suppose that the plotted line in Figure 9-4b will bend abruptly or change its angle, we can extend it to the edges of the diagram. We are using the same sort of reasoning (called *extrapolation*) that Galileo used when he said that if all resisting forces were removed, his blocks would slide along forever. Even though no one has set up a pendulum at the North Pole, we are sure that at the Pole (1.0 on the vertical scale of Fig. 9-4b) the pendulum's line of swing rotates at 15° each hour, or 360° each day. This is also the observed angular speed of each fixed star; therefore, at the North (or South) Pole, the lines AA', BB', and CC' in Figure 9-3 would all point to the same star; the pendulum's swing would "follow" a star.

Perhaps you have been wondering whether Foucault drew the wrong conclusion. Maybe Newton's law, $F = ma$, does not apply here or is wrong. Maybe the line of swing of the pendulum does actually rotate. But then, wouldn't it be an odd coincidence that at the North Pole it happens to rotate at exactly the same angular speed as the celestial sphere? (It would have to, since the line of swing of a pendulum there continues to line up with the same star.) If, on the other hand, you accept Newton's laws, you must then accept the reality of the earth's rotation. For if the line of swing remains the same (as his laws predict), and the celestial sphere turns, then the line of swing at the North Pole couldn't continue to line up with the same star. The turning of the sphere would carry a star away from the line of sight along the pendulum swing.

What happens at the equator gives further evidence that it is the earth that is rotating, and not the celestial sphere. There, as everywhere else on earth, each star appears to move in a circle around the earth once every 24 hours. But Figure

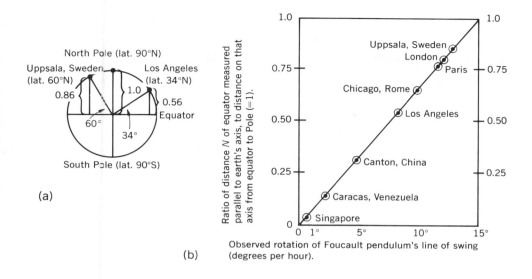

Fig. 9-4 (a) The diagram shows how the figures on the vertical dimension of (b) were obtained. Los Angeles is plotted at its latitude, 34° N. The distance of the North Pole from the equator (along the earth's axis) is taken as 1. Then the distance of Los Angeles above the equator, on a line parallel to the earth's axis, can be measured as 0.56. In a similar way, Uppsala, Sweden, at latitude 60° N can be measured as 0.86. (b)The vertical dimension shows the data described in (a). The horizontal dimension shows the angular distance through which the plane of swing of a Foucault penculum appears to rotate in an hour (8° in Los Angeles and 13° in Uppsala, for instance). The dot for Los Angeles is placed 0.56 above the base of the diagram and at about 8° from the left edge; that for Uppsala, 0.86 above the base and at about 13° from the left edge.

9-4b indicates that a pendulum suspended over the equator doesn't change its direction of swing with respect to compass points; extrapolation shows its rate of observed rotation to be 0° per hour. At the North Pole, a building and the place in it where the pendulum is attached are not carried anywhere by the earth's rotation but rotate through 360° each 24 hours. But if you were to look down from space directly on the equator, you would see the building and the pendulum move past you from west to east as the earth rotates, just as if they were being carried along by a freight train on a straight track. If the pendulum is started swinging parallel to a wall of the building, the wall will remain parallel to the line of swing. The line of swing will neither change direction nor appear to. This is just what we would predict if the earth rotates. But if the celestial sphere is doing the rotating, then you must explain why the plane of swing of a pendulum would rotate through 360° each day at the Poles and not at all at the equator.

If the earth rotates, everywhere between the equator and the Poles there is a combination of west-east motion (the building moving eastward) and circular motion (the wall of the building changing its direction with respect to the pen-

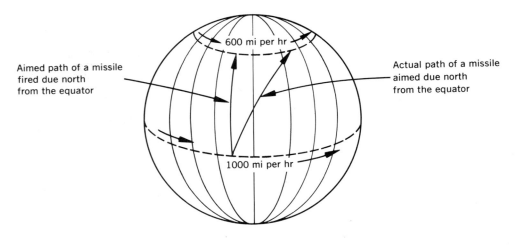

Aimed path of a missile fired due north from the equator

600 mi per hr

Actual path of a missile aimed due north from the equator

1000 mi per hr

Fig. 9-5

dulum's line of swing). The combination of these two kinds of motion causes the different hourly rates of apparent rotation of the pendulum's line of swing.

You can check to see that twisting the wire of Foucault's pendulum will not change the direction of swing. Hang a key or some other weight on a thread. Start it swinging, then twist the thread slightly, and you will find, as in the discussion above, that the direction of swing is not changed.

By the twentieth century, long-range artillery and missiles were furnishing a simpler indication that the earth rotates. Missiles launched in the northern hemisphere always veer to the east, and this must be taken into account when they are aimed. This can be explained in terms of Newton's laws and a rotating earth. If a missile is fired due north from the equator, it starts its trip with the northward velocity given it when it was launched. But it also starts the trip with an eastward velocity of about 1000 miles per hour, the velocity it shared with the rotating earth before launch. According to Newton, once the missile is launched, the only force acting on it is gravity, which acts in a downward direction only. Neither the northward nor the eastward velocity of the missile is affected (except for the slight effect of air resistance). As the missile moves northward, it passes over areas which are increasingly closer to the earth's axis of rotation (Fig. 9-5). Therefore, they do not move as fast as an area at the equator in completing their daily trip around the earth's axis. The farther north the missile goes, the greater is the difference between its "inherited" eastward velocity of 1000 miles per hour and that of the earth under it. So, relative to the ground, the missile veers to the east.

Additional Reading

ASHFORD, T. A., *From Atoms to Stars*: New York, Holt, Rinehart and Winston, 1960, pp. 37-38, 467-471.

COLEMAN, J. A., *Early Theories of the Universe*: New York, New American Library (Signet Science Library), 1967, Chap. 16.

chapter 10 | Masses in the Solar System

While we have been talking about mass so much, you may have noticed that we have never defined it, except rather vaguely as the total amount of material in a body. The more of it there is (larger m), the less it will be accelerated by a given force F. Mass is defined by Newton's equation $F = ma$. Mass isn't the same thing as weight; the weightless astronauts orbiting the earth still look the same; their masses have not changed.

Your weight is the gravitational force between the earth and you, and can be given by the familiar formula, weight $\propto mM_E/R^2$, where m is your mass, M_E is the mass of the earth, and R is your distance from the center of the earth. For everything at or near earth's surface, R is just about the same (4000 mi) and the mass of the earth is the same. Therefore, you can see that your weight depends on your mass. When you reduce your mass (by dieting, for instance), you reduce your weight correspondingly.

The whole earth is pulling on you. Why is the distance to the *center* of the earth used as R in the formula that determines your weight?

The earth consists of an extremely large number of particles. It is difficult to say what is meant by the word *particle* – rock masses, the crystals or grains in these rocks, the molecules in these crystals, or even the atoms in these molecules. Each of these particles is at a different distance from the earth's surface. Yet Newton's law of gravitation says that each of them is attracting an object near the earth's surface, pulling on it from many different directions and distances.

What is the result? Newton saw the apple fall straight down toward the center of the earth, along a line perpendicular to the earth's curving surface. This is the way everything falls to earth. Newton had to prove, however, that this path is predicted mathematically by his theory. He had to calculate the forces between an object at the surface of the earth and every possible particle of the earth, and then consider how all these forces combine. (To solve the problem, he had to invent a "new math," called the calculus, which has proved to be extremely useful in solving many other problems.) The result of all his calculations predicts what we observe: As far as gravitation is concerned, a spherical object such as the apple, the earth, the moon, or the sun, acts as though all of its mass were concentrated at its center. Each member of the solar system, regardless of its size, can be considered as a point where all its mass is concentrated. This makes exploration of the universe much simpler.

Thus, if you were to double your distance from the center of the earth, if you were to go 4000 miles above earth's surface and stand on a spring scale there, your weight would be cut to one-fourth of what it was. You would look the same (your mass has not changed), but you would not weigh as much. If you went to a planet the same size as the earth, but with twice the mass, your weight would be doubled. If you went to the moon, with r a little over one-fourth that of the earth and mass $1/80$ that of the earth, you would cut your weight to one-sixth of what it

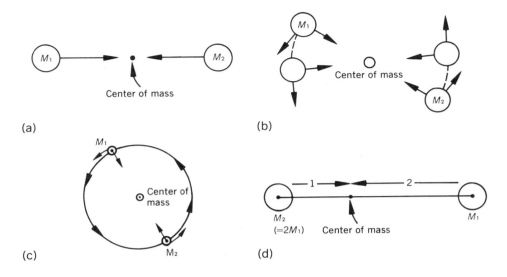

(a)

(b)

(c)

(d)

Fig. 10-1

was on earth. Careful measurement shows that you weigh less on the top of a high mountain, or even on top of a high stepladder.

As the earth pulls the apple downward, the apple pulls the earth upward, according to Newton. These two forces are opposite but equal in amount, that amount being the force of gravity between the two masses. They can be written as $\overrightarrow{F}_{EA} = \overleftarrow{F}_{AE}$. In other words, the force of the earth on the apple equals the force of the apple on the earth. Because the earth is so much more massive than the apple, the earth's upward acceleration is too tiny to be measured. What do Newton's laws predict if the two attracting objects are of equal mass? According to $F = ma$, they will be accelerated equally toward each other, since F and m are the same for both. They will collide at a point midway between their centers. This point is called their *center of mass* (Fig. 10-1a).

Now let us suppose that, instead of merely falling, these two equal masses, M_1 and M_2, have some sideways speed (Fig. 10-1b). M_1 is continually falling toward M_2, and M_2 is continually falling toward M_1. But both are moving sideways as well. The result is that they revolve around their center of mass, one opposite to the other If they are started with the correct sideways motion, they will move in the same direction around the same circle, always on opposite sides of the center of mass (Fig. 10-1c).

If one of these bodies is more massive than the other, the center of mass will not be midway between them. Suppose M_2 has twice the mass of M_1; then the center of mass is exactly twice as far from M_1 as from M_2 (Fig. 10-1d), and M_1 would follow a circular orbit twice as large as M_2's circular orbit. The center of mass is always nearer the larger of two masses.

If M_2 is much larger than M_1, its orbit becomes much smaller, and M_1 must go much faster than M_2 in order to stay on the opposite side of the center of mass. If we consider the sizes of the two bodies, it is possible that the orbit of M_2's center might be smaller than M_2 itself — that the radius of the orbit might be smaller than the radius of M_2. Then it would look as though M_1 were orbiting around M_2, although in fact both are circling their common center of mass. Only a very slow wobbling motion of the larger mass would reveal that it is moving in a tiny orbit.

Do we find any motion like this in the solar system? Yes, and right in our own back yard. As viewed from the earth, Mars doesn't appear to move among the stars at quite the speeds predicted. During the course of each month, Mars first appears to gain speed, so that it gets as much as 17″ of arc (almost $1/250$ of a degree) ahead of where it should be in the sky. Later it appears to slow down until it gets 17″ behind where it should be. The lead and lag repeats itself each month.

The most reasonable explanation is that as the center of the moon moves around the earth in a large orbit each month, the center of the earth moves around a tiny orbit. The center of both of these orbits is the center of mass of the earth-moon system. It is this center of mass that smoothly travels around the sun in a yearly orbit. But during this time the earth and the moon revolve around their common center of mass more than 12 times. If the earth is moving in a small monthly orbit, the surface of the earth (from which we view Mars) would sometimes get ahead of the center of mass of the earth-moon system and sometimes get behind that center.

Mars is shown in Figure 10-2a at the position where it is closest to the earth — about 35 million miles away. This is when the lead or lag is 17″. If angle a is only 17″, the distance of the center of the earth from the earth-moon center of mass can be measured as 3000 miles (Fig. 10-2b). The moon travels in a larger orbit with a radius of about 240,000 miles. Since the radius of the earth is about 4000 miles, the earth-moon center of mass is about 1000 miles below the surface of the earth. The earth merely wobbles around this center of mass. Venus and the other planets show a similar effect, smaller for the more distant ones.

A motion predicted by Newton's laws has been observed: a wobble in the earth's orbit due to the moon's pull, first one way, then another. The earth does not go smoothly around the sun, but the center of mass of the earth-moon system does. Because the earth is much more massive than the moon, the center of mass is inside the earth. Thus, the earth wobbles as the moon circles it, both moving in a yearly orbit around the sun.

The sizes of these monthly orbits of the earth and the moon enable us to calculate the relative masses of the earth and the moon. The center of the earth is 3000 miles from the center of mass. The center of the moon is 240,000 miles from the center of mass: 1000 miles (radius of the moon) + 238,000 miles (distance from the surface of the moon to the surface of the earth) + 1000 miles (distance from the surface of the earth to the center of mass) = 240,000 miles from the system's center of mass. The center of the moon is $240{,}000/3{,}000$ times farther from the common center of mass than is the earth's center. Therefore, the mass of the moon is about $1/80$ that of the earth.

Masses in the Solar System

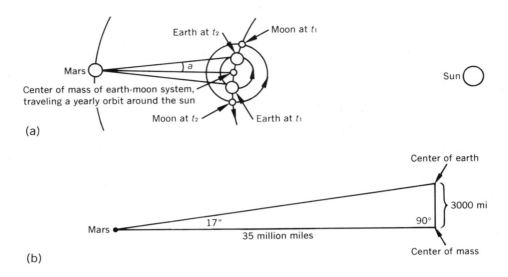

(a)

(b)

Fig. 10-2 (a) If the earth traveled the orbit shown here, around the center of mass of the earth-moon system, the line of sight from earth to Mars would vary by angle *a* from the line of sight to Mars from the center of mass. Mars would be observed first ahead and then behind its predicted path among the stars. (b) Angle *a* is observed to be 17″ when Mars is 35,000,000 miles away. Therefore, the center of the earth is 3000 miles from the center of mass of the earth-moon system.

Newton's laws predict that the sun's center is not the center of the orbit traveled by the earth-moon system each year. We would expect this system and the center of the sun to travel around their center of mass, the sun moving in a small orbit while the center of the earth-moon system travels in a larger one. Each planet and the sun should move in a pair of orbits; and so the sun must be pulled every which way as the planets move around it. The sun should "wobble" too, but even the most precise instruments have never detected it. We must conclude that the sun's mass is so large compared to that of the planets that the wobble is too small to be observed.

Is there any way in which we can use Newton's laws to determine how much greater the sun's mass is than the earth's mass? Can we determine the mass of the earth or of the sun in pounds, or tons, or grams?

We have written Newton's law of gravitation as $F \propto m_1 m_2 / R^2$. This is like saying that your bill at the doughnut shop will be proportional to the number of doughnuts you buy (bill ∝ number of doughnuts bought). You know that it will cost ten times as much to buy ten doughnuts as it does to buy one, even though you don't know how much in dollars and cents. However, if you buy one doughnut and are charged $.05, then you know that the bill for ten doughnuts will be $.50. You can now write the equation: Bill = $.05 × Number of doughnuts bought. The cost per doughnut is called a *proportionality constant*. In order to replace the proportion sign (∝) with an equals sign (=) you must include the proportionality constant in the equation. In the same way, before we can replace the $\propto$ in Newton's law with $=$, the missing constant must be supplied.

This was done by Henry Cavendish, an English scientist, in 1798. He connected two metal spheres by a very light rod and suspended the rod by a thin thread at the middle. Then he placed a large mass beside one of the spheres and another large mass beside the other. The gravitational forces between these masses and the two spheres caused the spheres on the rod to twist the thread. The amount of twist in the thread measured the force between the mass of the sphere and the larger mass. Cavendish had measured the masses in pounds (or grams) and the distance between them in inches (or centimeters). Therefore, he could calculate from Newton's law what the force would have been if the spheres and the large masses had each weighed one gram and if they had been one centimeter apart. If this force had proved to be one dyne (the force necessary to accelerate a mass of one gram to a speed of 1 centimeter per second in a second), then the $\propto$ in Newton's equation could have been replaced by an equals sign. But the force was much smaller. It was only 0.0000000668 dyne, or as it is usually written, 6.68×10^{-8} (Calculation 1). This is the gravitational force between two 1-gram masses which are one centimeter apart. It plays the same role in the equation as the $.05 per doughnut did in the bill-for-doughnuts equation. It is the constant of gravitation G and must be inserted in Newton's equation along with the equals sign: $F = G m_1 m_2 / R^2$.

Now, using Newton's equations, we can determine the mass of the earth, as shown in Calculation 2. It turns out to be 6.6×10^{21}, or 66 followed by 20 zeros tons (Calculation 1). The mass of the moon is thus $6.6 \times 10^{21}/80 = 8.25 \times 10^{19}$ tons.

Back in the sixteenth century, even though the sun's diameter had been estimated to be 600 times that of the earth, it did not seem odd to consider this larger body as circling the earth. By 1800, the size of the sun had been quite accurately measured, and Newton's laws had explained the falling apple and the orbiting moon. With larger telescopes, more observatories, and better communication, the parallax of Mars, whose orbit lies only about $1/2$ AU from earth's (Table 1), could easily be determined. The parallax, observed simultaneously from widely separated places on earth (whose distances apart were accurately known), showed Mars to be a little over 46.5 million miles from us. Then, 0.5 AU equals 46.5 million miles, and our distance from the sun (1 AU) is about 93 million miles. The angular diameter of the sun is one-half of a degree. An object that is one-half of a degree wide at a distance of 93 million miles can be shown by geometry to be about 865,000 miles in diameter. The earth's diameter is 8000 miles. The volume of a sphere varies as the cube of the radius. Hence the volume of the sun is more than a million times the volume of the earth, since its radius is a hundred times that of the earth. It was immediately seen that the sun could be made of material thousands of times lighter than that of the earth and still be much more massive than the earth. As more and more predictions from Newton's laws were made and proved, this came to be taken as strong evidence for the acceleration of the earth in an orbit around the sun.

Masses in the Solar System

We were able (Calculation 2) to compute the mass of the earth using the acceleration of a body falling toward the earth. We should also be able to calculate the mass of the sun because the planets are falling toward the sun. This is a bit more complicated, because the planets are traveling in orbits around the sun. Calculation 3, based on Newton's laws and following the rules of algebra and geometry, shows that the mass of the sun is given by the equation $M = 4\pi^2 R^3 / P^2 G$, where R is the distance of the earth from the sun, P is the earth's period, and G is the constant of gravitation. The mass of the sun turns out to be 333,000 times the mass of the earth. No wonder that the yearly wobble of the sun cannot be detected! It would be 300 miles: 93,000,000 miles/333,000 = 300 miles.

In the equation $M = 4\pi^2 R^3 / P^2 G$, the quantities $4\pi^2$, G, and M stay the same, whatever planet is used to measure the sun's mass. Kepler found that for all the planets, $P^2 = R^3$. Therefore, we can use any planet to determine the sun's mass.

And, of course, we can use the equation $M = 4\pi^2 R^3 / P^2 G$ to calculate the mass of the earth, by considering the moon to move around the center of the earth, neglecting the earth's wobble. In this calculation, R is the distance of the center of the moon from the center of the earth and P is the period of the moon ($27\frac{1}{2}$ days). The earth's mass comes out at about 6.6×10^{21} tons, just as it did when we calculated it by means of the falling apple.

In the same way, the mass of any planet can be determined if we know the period of any one of its satellites and the distance of that satellite from the center of its planet. One of Jupiter's moons (Fig. 10-3), called Io, is 262,000 miles from the center of Jupiter and has a period of 1.75 days. When this period (in seconds) and the distance (in centimeters) are put into the equation $M = 4\pi^2 R^3 / P^2 G$, we get $M =$ Jupiter's mass in grams. Jupiter's mass turns out to be 317 times earth's mass, or about $\frac{1}{1000}$ the mass of the sun. Any of Jupiter's moons will give the same result. The periods and distances of Saturn's moons show it to have 95 times the earth's mass.

The large masses of Jupiter and Saturn remind us that if "every particle of matter in the universe attracts every other particle," then the planets must attract each other. The amount of gravitational force between any two planets, of course, depends on their masses and the distance between them. Because the planets are traveling along their orbits around

Fig. 10-3 Venus, above, and Jupiter with four bright moons (the ones Galileo saw) photographed with a 10-inch telescope. [Yerkes Observatory photograph.]

the sun at different speeds, the distance between any two of them is constantly changing. In Figure 10-4, for instance, at time t_1 the gravitational force between Mars and Jupiter is greater than at time t_2. We observed that when Kepler plotted the positions of Mars in its orbit, these positions did not exactly fit an ellipse. When he replotted them, taking into account the eccentricity of the earth's orbit, they fit somewhat better. But records of Jupiter's positions in its orbit at the time of each observation of Mars show

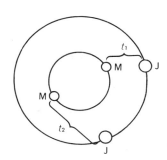

Fig. 10-4

that the deviations of Mars from its predicted positions can be explained. Jupiter was *perturbing* Mars, pulling Mars off its true elliptical orbit. The amount of the pull varied, being greatest when the two planets were closest.

Jupiter is perturbed by Mars too, but not to the extent that Mars is perturbed by Jupiter. The mass of Mars (6.7×10^{20} tons, or 0.107 of the earth's mass) was calculated from the perturbing effect of Jupiter. When the two moons of Mars were discovered in 1877, Mars' mass was calculated from their periods and distances from the planet's center, and it agreed with this figure. Venus has no moons; its mass (0.814 times that of the earth), is measured only by the perturbing effect it has on the earth. Venus' effects on Mercury's orbit show Mercury to have a mass 0.054 times that of the earth.

Of course, all the planets affect each other, and the results of the interaction of all these masses is a difficult mathematical problem. Modern electronic computers are a great aid in solving these problems and predicting planet orbits accurately.

Additional Reading

ABELL, G. O., *Exploration of the Universe:* New York, Holt, Rinehart and Winston, 1964, Chap. 4.

TRICKER, R. A. R., *The Paths of the Planets:* New York, American Elsevier Publishing Company, 1967, Chap. XIII.

chapter 11 | **Uranus, Neptune, and Pluto**

In 1757, a 19-year-old German musician named Friedrich Wilhelm Herschel fled to England to escape military service in the Seven Years War. A decade later, after a great deal of hard work, he was comfortably settled in the resort town of Bath, and everyone called him William Herschel. He was a capable and

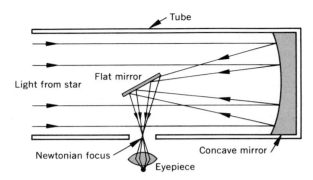

Fig. 11-1 The reflecting telescope. The magnification (increased separation of two close stars, or increased size of the moon or a planet) depends on the distance from primary mirror to focus (the focal length), and on the power of the eyepiece. A high-power eyepiece can be used close to the focus. The increase in brightness of a star image, over that seen with the naked eye, depends on the area of the primary mirror.

popular music teacher and director of the city's orchestra. In 1772, a friend lent him a telescope, and, as Herschel said, "this opened the kingdom of the skies" to him. He became an enthusiastic amateur astronomer. He also became a telescope maker, after finding that good ready-made telescopes were too expensive for his means. In his spare time Herschel tried first to make a refracting telescope, which uses lenses to produce a magnified image (Figs. 7-1 and 7-2), but he could not grind any lenses that were good enough to satisfy him.

Then he began to make a reflecting telescope, using mirrors instead of lenses. The first successful telescope of this type had been built by Sir Isaac Newton in 1668. In a reflecting telescope (Fig. 11-1), a concave mirror at the bottom of the tube takes the place of the lens at the top of the tube. This mirror reflects the light from a star back up the tube. An image of the star is formed near the front end of the tube, where the reflected rays come to a focus. This focus is in the path of the incoming light, of course, but Newton had solved the problem by placing a flat mirror, mounted diagonally, in the middle of the tube, as shown in Figure 11-1. This mirror intercepts the light just before it reaches the focus and reflects it to the side, so that it comes to a focus outside the telescope tube and can be viewed through a small lens (the eyepiece).

Until Herschel made them popular, few reflecting telescopes were used. Good telescope mirrors are difficult to make. Before Herschel produced a satisfactory one, he made and discarded about 200 of them. But after a few years' work he produced a really excellent instrument, 7 feet long, with a mirror 6½ inches in diameter. Using this telescope, Herschel began to map the sky methodically, piece by piece, constellation by constellation, including every star he could see. He measured and plotted their angular distances from each other, exact to a few seconds of arc. He was helped in this work, and indeed throughout his career, by his sister Caroline, the first woman astronomer. She did many tedious and difficult calculations for him and used the telescope from time to time as well.

By March 1781, Herschel had reached the zodiac constellation Gemini (The Twins) in his survey of the sky. On the night of March 13, he was studying a star field in Gemini. About 10:00 his eye was caught by a star brighter than the others in the field of view of his telescope. He was surprised to find that it was not shown on the older, less detailed star maps and lists that he was using as a guide. Also,

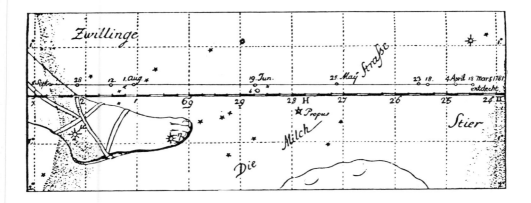

Fig. 11-2 Johann Bode's diagram of the apparent path of Herschel's "comet" from observations between March 13 and September 13, 1781 (the line with the small circles and dates). The blocked line parallel to it and slightly south is the ecliptic (the apparent eastward path of the sun around the celestial sphere). "Die Zwillinge" is the German word for "The Twins"; "Die Milch Strasse" is the Milky Way. [From *Berliner Astronomisches Jahrbuch,* 1784.]

it appeared to be a disk, rather than a point of light. He thought that it probably was a comet (Fig. 5-2). The eyepiece that he was using magnified images 227 times. He exchanged this eyepiece for a stronger one, increasing the magnification to 460 times. The diameter of his "comet" image doubled, while the images of stars still looked like points of light (as they should in a good telescope). He changed eyepieces again, and now the images were magnified 932 times. The star images were little changed, but the "comet" seemed twice again as large. Now he was certain that it was not a star.

He watched the "comet" for a month and saw that it moved relative to the fixed stars. In early April he sent notices of his discovery to astronomers all over Europe, asking them to track it. Soon it was being observed nightly by many skilled astronomers using the best telescopes in the world. Among them was a German astronomer named Johann Bode at Berlin Observatory, whose map of its movement in the sky is shown in Figure 11-2.

By the end of May the sun had moved so close to the "comet" on the celestial sphere that the "comet" was visible only for a few minutes after the sun went down. It couldn't be seen during most of June, and then, late in the month, it appeared in the sky just before sunrise (Fig. 11-3). If it were a comet, traveling in a long, oval orbit around the sun (Fig. 11-3d), this was the very time that it should have developed a tail. (From ancient times it had been noticed that all comets develop a tail as they approach the sun.) But no tail was observed. Furthermore, it looked different from any known comet in other respects: It was not fuzzy around the edges; it was a clear, sharp disk like Jupiter or Venus. Although its eastward movement among the stars increased as it neared the sun, it didn't increase nearly as much as a comet's would have. (In accordance with Newton's laws, note how much a comet's speed would increase between *a* and *c* in Figure

77

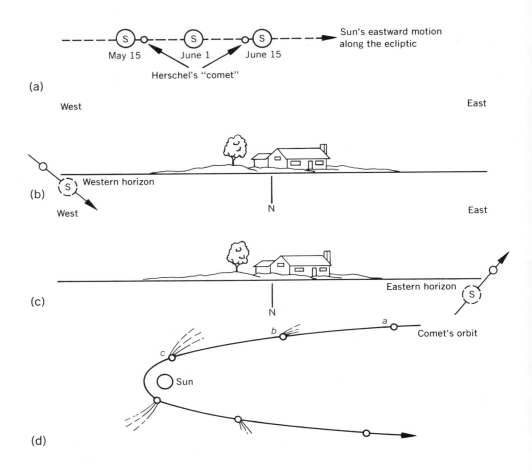

Fig. 11-3 (a) Relation of Herschel's "comet" to the sun, May 15 to June 15, 1781, showing eastward motion of both along the ecliptic. (b) On May 15 the "comet" was visible for a short while after the sun had set in the northwest. (c) On June 15 the "comet" rose shortly before the sun did and was visible for a short time before its light was drowned by the sun. (d) Development of a tail as a real comet approaches the sun; disappearance of the tail as the comet recedes from the sun.

11-3d.) In fact, as Figure 11-3a shows, it seemed to move eastward more slowly than the sun. Furthermore, there was no change in its apparent diameter, as there would have been if it had been traveling in an orbit like that in Figure 11-3d.

By November 1781, part of its orbit around the sun could be plotted, in the same way as Copernicus plotted the planet orbits (Figs. 4-5 and 4-6). It turned out to be part of a circle with a radius of 19 AU. The full orbit would completely encircle that of Saturn; and Herschel's new planet, as it now appeared to be, was double Saturn's distance from the sun (9.5 AU). Kepler's harmonic law $(P^2 = R^3)$

indicated that it would take 80 years to travel once around this orbit. Its distance and its angular diameter in the sky (5″) showed that this new planet's diameter was more than 3½ times that of the earth.

Herschel wanted to name his planet Georgium Sidus (Latin for "The Georgian World") in honor of George III, then King of England. Such a name would have made it as popular as a hornet in the American Colonies, then in the midst of the Revolutionary War. However, Bode's suggested name, *Uranus,* was accepted. In Greek mythology, Uranus was the father of Saturn, who was the father of Jupiter.

The discovery of Uranus wakened a popular interest in astronomy that was never equalled until after Sputnik was launched in 1957. Social and official London rushed to do Herschel honor; his front yard was crowded with visitors, and the way to his observatory was blocked by carriages filled with curious and admiring throngs. Terms like *orbit* and *angular diameter* were heard for the first time in many a drawing room and pub, and everyone wanted a look at the new planet. George III appointed Herschel as his Court Astronomer, and he was able to devote himself entirely to astronomy. In 1787 he discovered two moons of Uranus, and this enabled the mass of the planet to be calculated: It is 14½ times that of the earth.

Uranus can often be seen without a telescope on a clear, dark night, and astronomers of Ptolemy's day and before must have seen it. However, its motion is so slow, and it is so faint, that they did not recognize it as a "wanderer." In the widespread rush of interest in the new planet, old star charts were brought out, dusted off, and looked over. Uranus' position in the sky at the time a chart was made was determined (from the planet's known period and orbit). Then the charts were compared with modern ones. On 23 of the old charts an extra "star" was found at the predicted position of Uranus. The first observation had been recorded by Tycho, and the remaining 22 were telescopic observations, the oldest made in 1690. Uranus had been seen in six different constellations of the zodiac and mistaken for a star during the 90 years prior to its discovery.

As soon as an orbit had been worked out for Uranus, tables were computed to show the position of the planet in the sky for succeeding years. Of course, the perturbing effects of the other planets were taken into account. In 1798 the observed positions were "off" from the predicted ones only by 1″ of arc (1/3600°); this was not serious. But by 1810 they were off 6″ (well above the possible error of a skilled observer with a good telescope) and the differences were increasing. In 1821, a French astronomer, Alexis Bouvard, decided to rework the orbit calculations, in the hope of making more accurate tables. He had available 22 prediscovery positions covering 90 years, and nightly records of the planet's positions for the 40 years since 1781. To his surprise, he found that he couldn't construct an orbit which fit them all. When he included the prediscovery positions, the present position of Uranus was 45″ off. If he used only the postdiscovery observations, the present position was only 9″ off. He decided to discard the 22 older observations, which were, he reasoned, more likely to be in error. He drew up his tables based on an orbit corrected to take care of the 9″ discrepancy.

But Uranus didn't stick to this orbit. In 11 years most of the positions in Bouvard's tables were in error by ½' (30″) and the tables had to be constantly revised, just as in the days when Ptolemy's tables were used. By 1844 the discrepancy had increased to 2' (about one-fifth of the moon's diameter).

If the planets were behaving according to the law of gravitation, there should have been no such differences between predicted and observed positions. It was generally believed that Newton's laws were universal, or that they, at least, held to the limits of the solar system. If they did not, then exploration of the universe beyond Uranus, out in the realm of the fixed stars, would be difficult without his laws as guides.

Astronomers realized that gravitational attraction of another planet beyond Uranus could be disturbing its motion, making it appear to be moving contrary to Newton's laws. In 1841, John C. Adams, a 22-year-old student at Cambridge University, decided to compute where such a planet must be and what orbit it would follow in order to produce the unexplained parts of Uranus' motion.

The prediscovery positions of Uranus (with the effects of perturbations by the known planets subtracted) can all be plotted on an almost circular orbit. This orbit, however, does not fit the postdiscovery positions. In Figure 11-4, the inner circle represents the orbit followed by Uranus during the 91 years from 1690 to 1781. Several observed positions of Uranus during the following 60 years (1782-1842) are shown by dots. The predicted positions on the basis of this orbit are shown by open circles. (The differences between observed and predicted positions are exaggerated to make the diagram legible.) It seems reasonable that any "new" planet, which we may call Planet X, was farther from Uranus during the time when the steady orbit

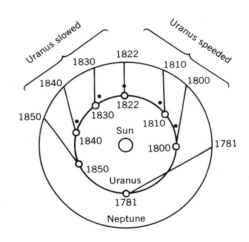

Fig. 11-4

was followed more closely because then the perturbing force would be less, according to Newton's formula, $F = Gm_1m_2/R^2$.

Adams first examined the positions of Uranus from 1782 to 1822 (Fig. 11-4) and concluded that during this time the faster moving Uranus was catching up with Planet X. As they came closer, the gravitational force between them increased, and Uranus was pulled farther from its true orbit. During this time, Uranus was speeding up and getting ahead of its predicted positions. This meant that Planet X lay somewhat ahead of Uranus and pulled it forward (Fig. 11-4). From 1822 to 1842, however, Uranus' observed positions were coming closer to the predicted ones; that is, Uranus was slowing down. Adams concluded that during this period Uranus had overtaken Planet X and was now ahead and was being pulled backward. In 1822 the difference between observed and predicted positions was greatest, so it was probable that the two planets had passed each other then.

Adams reasoned that the relative positions of the two planets from 1782 to 1842 were something like those shown in Figure 11-4. But were the relative sizes of the two orbits anywhere near right? How far apart were Uranus and Planet X at each date? By measuring the perturbations of Uranus caused by Planet X, he could determine the gravitational attraction between them on any date and at each of the positions, for instance, shown in Figure 11-4. According to Newton, this force F is equal to Gm_1m_2/R^2, where $m_1 =$ the mass of Uranus, $m_2 =$ the mass of Planet X, and $R =$ the distance between them.

There were two unknown quantities in this equation, m_2 (which, of course, must stay the same) and R (which is continually changing). As Adams made m_2 larger, R^2 was larger for each position and date. The size of each R, the distance of Planet X from Uranus on each observation date, would give a series of points that lie on the orbit of Planet X. They would give R, the planet's distance from the sun.

Adams tried many combinations of m_2 and R, and then tested the resulting orbits with the following considerations in mind: (1) Since the size of Uranus' orbit was known, the timing of the variation in perturbation (Fig. 11-4) showed him how fast Planet X must be moving along its orbit. Then from Kepler's law, $P^2 = R^3$, he could determine the radius of the orbit. (2) He may also have been influenced by a curious mathematical relation between the distances of successive planets from the sun. This relationship, publicized by Johann Bode in 1766, predicted that a planet beyond Uranus would be about 39 AU from the sun. (3) It was unlikely that the mass of Planet X would be smaller than that of Venus or larger than that of Jupiter, but it had to be large enough to pull Uranus the right amount from the distances he had calculated.

By October 1845 Adams had a solution that satisfied him. He predicted that Planet X's orbit would be 36 AU from the sun and that its mass would be about 25 times that of the earth. He also calculated where a planet following this orbit would be found in the sky during the next few years. He took his computations and predictions to Professor J. Challis at Cambridge, hoping that Challis, an astronomer, would look for Planet X with the university's telescope. Challis was not interested, and sent Adams to Greenwich to see the Astronomer Royal, who was not interested either.

Meanwhile, Urbain J. J. Leverrier, a French mathematician somewhat older than Adams, had tackled the problem without knowing of Adams' work. In the summer of 1846 he presented three papers to the French Academy on the subject. Now the Astronomer Royal pricked up his ears, and ordered Challis to look for the planet in the position where Adams predicted it would be. Challis made a map of the stars in that position on August 4 and another on August 11. When they were compared, he found no planet. Yet later inspection of his maps showed that he had not compared them carefully enough. One star had moved, and it was Planet X.

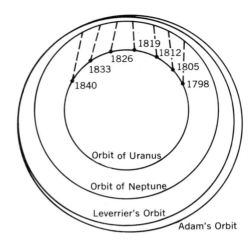

Fig. 11-5 The orbits of Uranus and Neptune, and Adams' and Leverrier's predicted orbits for Neptune. [Courtesy of Harvard College Observatory.]

On August 31, Leverrier finished his calculations and sent his estimate of Planet X's position in the sky (in the constellation Aquarius) to the Berlin Observatory, where he knew that charts of that portion of the sky had recently been made. J. G. Galle, an observer there, received his letter on September 23 and that very evening turned his telescope to the place in the sky directed by Leverrier. Although no object in the telescope field showed as a disk, there were nine stars there and only eight on the chart! Was one of them the planet? The next night Galle found that this ninth star had moved slightly with respect to the others. The new planet had been found! As Adams' predicted position was almost as accurate as Leverrier's, both men are given credit for discovering Neptune, as the planet was later named.

The discovery of Neptune was another triumph for Newton's theory, and a reassurance that it could be used in further exploration of the universe. In time, the orbit of the planet was computed and found not to be exactly as predicted by either Adams or Leverrier (Fig. 11-5). It is 30 AU from the sun, with a period of about 165 years. Its angular diameter (about 3″) indicates that it is about the same size as Uranus, and its mass was later determined from the motions of its moons to be 17.6 times that of the earth.

Even after the effect of Neptune was taken into account, differences of 2″ remained between the predicted and observed positions of Uranus, and Neptune itself was not keeping exactly to its orbit. In 1906, an American astronomer, Percival Lowell, calculated that there was a planet beyond Neptune and about 45 AU from the sun. He predicted that this planet would appear small and faint, a little over 1″ of arc, and that it would spend about the next 30 years in the constellation Gemini. At his Arizona observatory, Lowell searched for the planet without success, until his death in 1916.

Such a faint and slow-moving object is much more difficult to find than Neptune, but by this time there were new aids to its discovery. In 1849, Henry Draper, an astronomer at Harvard College Observatory, first used the telescope as a large camera by placing photographic film at the telescope focus (Figs. 7-1 and 11-1). In this manner time exposures can be made, and the light falling on the film from a star gradually accumulates and builds up a photographic image. Since Draper's time, photographic films have been made more sensitive, and it is now possible to photograph objects which are a hundred times too faint to be seen by looking through the telescope.

Lowell's search had been made by studying photographs, and after his death others kept on looking for the planet. The search was intensified in 1929, when a new 13-inch refracting telescope was put into operation at Lowell Observatory. One-hour exposures were made on photographic plates which recorded an area of the sky 12° × 14°, at a scale of 1' = 1 millimeter (about 0.04 inches). The entire area of the sky around the zodiac was photographed, and a photograph of the same region was taken three times, usually within a week. Each plate recorded from 5000 to 400,000 stars.

At first glance, the task of comparing plates of the same area to see if any stars have moved appears hopeless. However, an invention called the blink microscope makes it possible. Two photographs of the same area, taken on different nights, are placed side by side. The operator's view through a lens is automatically shifted back and forth between corresponding parts of the two photographs, which are carefully adjusted so that fixed stars appear in the same place either way. If one object has moved relative to the stars in the interval between the taking of the two photographs, the image of that object appears to jump back and forth. In this way, any object that is moving can be picked out quickly and easily.

In February 1930, Clyde Tombaugh, a 24-year-old astronomer who had recently arrived at Lowell Observatory, was comparing photographs made on January 23 and January 29 of that year. In the area where Lowell had predicted the new planet would be found, he saw an object whose motion appeared about right for one beyond Neptune's orbit. It was within 6° of Lowell's predicted position, in the constellation Gemini. Additional photographs were taken as soon as the weather permitted, and on several succeeding nights. They showed that the new planet was moving westward at about 70" per day. Its orbit proved to be very similar to the one which Lowell had predicted. Announcement of the discovery was made on March 13, the 149th anniversary of Herschel's discovery of Uranus. The planet was named Pluto, after the Greek god of the underworld (its first two letters are Percival Lowell's initials). It is 39.5 AU from the sun, its period is 248 years, and its mass is not yet determined with any certainty.

From 1930 on, a detailed examination of photographs already made and a new photographic survey of the region of the ecliptic has been going on at Lowell Observatory, to see whether another planet lies beyond Pluto. None has so far been found. The present picture of the distances, sizes, speeds, and masses of the planets is shown in Table 4.

	Mercury	Venus	Earth	Mars	Jupiter	Saturn	Uranus	Neptune	Pluto
Average distance from sun (in AU)	0.39	0.72	1.00	1.52	5.20	9.54	19	30	40
Average distance from sun (in miles)	3.6×10^7	6.7×10^7	9.3×10^7	1.42×10^8	4.83×10^8	8.86×10^8	1.78×10^9	2.79×10^9	3.67×10^9
Period	88 days	225 days	1 year	687 days	12 years	29.5 years	84 years	165 years	248 years
Diameter, in terms of earth's diameter (8000 mi)	0.37	0.96	1.00	0.52	10.9	9.1	3.7	3.5	0.5?
Number of moons observed	none	none	1	2	12	9	5	2	none
Mass, in terms of earth's mass ($=6.6 \times 10^{21}$ tons)	0.05	0.81	1.00	0.11	317	95	14.5	17.6	0.18
Weight of a person weighing 100 pounds on earth (in lbs)	33	88	100	41	267	115	106	144	?

Table 4 Data on the nine planets.

Additional Reading

LUBBOCK, C. A., *The Herschel Chronicle:* Cambridge, Mass., Cambridge University Press, 1933.

MICZAIKA, G. R., and W. M. SINTON, *Tools of the Astronomer:* Cambridge, Mass., Harvard University Press, 1961.

SPENCER-JONES, SIR HAROLD, "John Couch Adams and the Discovery of Neptune" in *Astronomy* (Samuel Rapport and Helen Wright, eds.): New York, New York University Press, 1964.

TOMBAUGH, C. W., "Reminiscences of the Discovery of Pluto" in *Sky and Telescope,* Sky Publishing Corp., Cambridge, Mass., March 1960.

chapter 12 | # Colors and the Spectrum of Light

Back in the days when chandeliers with glass prisms were fashionable, everyone knew that sunlight shining on these prisms made little rainbows of color on the wall. Up until Newton's time, people thought that the material of the glass itself produced these colors. Of course, they all saw sunlight coming through glass windowpanes without forming any rainbows. But very few of them wondered why the glass prism changes the light and the glass window does not.

Then in 1666 Newton let a narrow beam of sunlight shine on a prism and saw that each color of light was coming out of the prism at a slightly different angle, thus spreading the colors into a rainbow (Fig. 12-1). The red light was bent the least, the violet light the most.

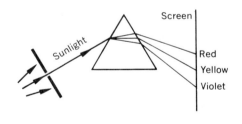

Fig. 12-1 Rainbow from a glass prism.

Then he put a second prism, upside down, a short distance away, as shown in Figure 12-2a. He saw that the light which came out of the second prism was white, like sunlight. If he moved the two prisms so that they touched (Fig. 12-2b), their two outer sides were parallel, as in window glass, and the light that came through was still white. From this experiment he concluded that the glass in the prism does not make any new kind of light; it simply separates white light into colors. This meant that white light is a mixture of all colors of light. If you put the colors back together, as Newton did with his second prism, you have white light.

85

Colors and the Spectrum of Light

But what causes the separation of white light into colors? Newton shone beams of sunlight through the thick and thin parts of a prism and saw that as long as he kept the beams parallel, the light of any one color always came out at the same angle. The distance that the light traveled through the glass apparently had nothing to do with the bending that separated the colors. It all happened at the air-glass surface. Did the bending take place where the light went in, or where it left the glass? He took a prism of a slightly different shape, one in which the far side of the prism made a different angle with the near side, and found that the light of each color came out at a different angle than before. This showed him that the light is bent as it enters the glass and again as it leaves the glass. Now he was able to explain how light shining through the two prisms could be white (Fig. 12-2b).

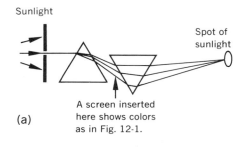

(a)

A screen inserted here shows colors as in Fig. 12-1.

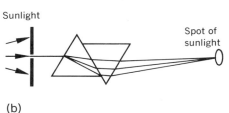

(b)

Fig. 12-2 Newton's experiments with prisms.

But what causes the light to bend when it goes from one sort of material into another, as from glass to air or from air to glass? Newton concluded that light is made up of very tiny particles. These particles, he said, form rays of light as they travel in straight lines at enormous speed. Particles of all colors of light, he believed, move at the same speed in a thin material like air. But they move at slower rates, different for each color, in a solid material like glass. Figure 12-3 shows that a change of speed of the particles could change the direction of a light ray.

Fig. 12-3 According to Newton's theory of light, as a ray of sunlight enters glass, the particles of light (shown here as lines) are slowed down. When a ray strikes the glass at an angle, the part of a line striking the glass first will not travel as fast as the part of the line still in the air. This second part will then pivot around the first part until it, too, has sunk into the glass. The pivoting done by each succeeding line causes the ray of light to travel in a new direction through the glass. Each line in turn changes direction in the same way. When the ray of light emerges from the other side of the glass, one portion of each line will travel faster than the part remaining in the glass, and the reverse of what happened when the ray entered the glass will take place. To make the diagram simpler, only one color of light is shown.

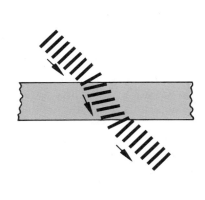

Light from a candle spreads out, lighting the room. The light of the sun, 93 million miles away, reaches the earth, lighting and warming it. According to Newton, the light is shot out from its source like a continuous stream of bullets. Christian Huygens, a Dutch scientist of Newton's day, suggested another way that light could be traveling. He pointed out that when a stone is dropped in a calm pond, circular waves move outward, away from the place where the stone landed in the water. The waves reach the edge of the pond and batter the shore, far from where the stone was dropped. As long as you continue dropping stones in the pond, waves continue to move outward. Huygens pictured waves of light spreading out in all directions from a lighted candle or from the sun.

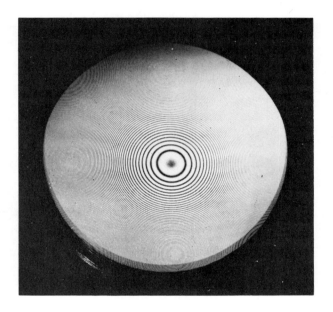

Fig. 12-4 "Newton's rings." [Courtesy of Bausch & Lomb, Rochester, New York.]

A long controversy developed over these two explanations of light. There really are no other sensible ways of explaining how light gets from its source to a distant place where its effects can be seen or felt (or photographed nowadays). Such "actions at a distance" could only be explained by particles or waves. A brighter light would be caused by more numerous particles or higher waves; different colors of light could be different sizes of particles or waves. That is, both ideas could explain most of the things that light does. But one of Newton's experiments could be explained by the wave theory and not by the particle theory. This was "Newton's rings" (Fig. 12-4), formed where light passes across a fine gap between two glass surfaces. In fact, it was this experiment reported by Newton that suggested the idea of the wave theory of light to Huygens.

The waves that he was familiar with are waves on water—on the sea or in a pond or lake. In a water wave, the water itself does not move forward with the

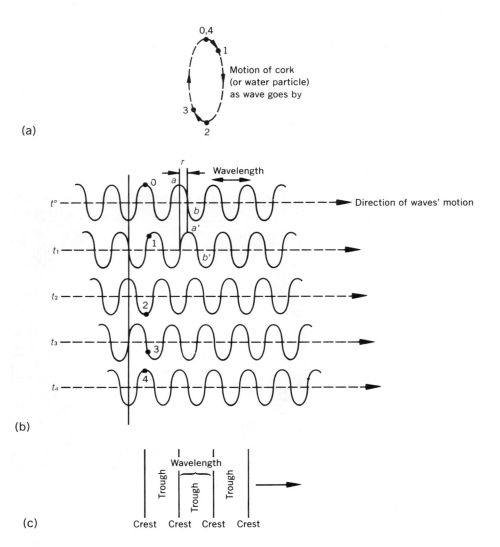

Fig. 12-5 (a) Motion of a cork (or water particle) as a wave goes by. A tiny ellipse is described. (b) The motion of the water particles in ellipses produces waves. Positions *0, 1, 2, 3,* and *4* of the particles are the same positions as in (a). The wave forms move in the direction of the arrow. (c) The appearance of the waves in the last line of (b) as viewed from above. The lines show the wave crests moving in the direction of the arrow.

wave. You can see this by watching a cork floating on the surface of the water. It bobs up and down a little and forward and back a little, moving in a small ellipse (Fig. 12-5a). The cork is doing what each water particle is doing. All of them are moving in ellipses, each one set off by the motion of the one next to it, like the diagram in Figure 12-5b. The normal (unwaved) surface of the water is shown by the dashed lines. The crest of one wave is at a, where the particles are highest above normal. The trough of the wave is at b, where they are farthest below normal. In just a moment, however, the motion of each water particle has caused the crest to move to a', and the trough to b'. If the distance between a and a' (or b and b') is r inches, and the time between the two drawings is t seconds, then the speed of the wave is r/t inches per second.

Wavelength is the distance from one crest to the next (or from one trough to the next), as shown in Figure 12-5b. If you were to look down on the waves in the bottom row of Figure 12-5b (which is a cross section), you would see each advancing wave crest forming a line, as shown in Figure 12-5c. The distance between the lines formed by the crests is the wavelength.

When waves on the water move into shallower water, they are slowed down. You can often see waves coming in at a slant to the coast, as shown in Figure 12-6a. The lines a through e are the crests of waves approaching the shallower water. Let us look at crestline b. The part of it north of the shallow-water area is still moving along at its old speed; but the southern part of the wave, which has already reached the shallow water, has been slowed down. The same thing is happening to wave crests c, d, and e, and it has already happened to wave crest a. Not only has the speed of the waves become less, but the direction in which they are moving (shown by the arrows) has changed.

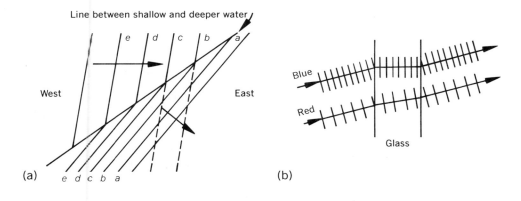

Fig. 12-6

Figure 12-6b shows how the wave theory explains what happens to white light when it enters a prism. The white light is a mix of all colors. Just two are shown: long red waves and short blue ones. The arrows show the direction of the incoming beam of light (this is the direction in which all the waves are moving). Notice that the blue light is bent more than the red. Therefore, it must be slowed down more by the glass than the red light is.

According to the wave theory, when light strikes a glass (like a windowpane) at an angle, all the waves are slowed down. But when they come out of the glass, the

Colors and the Spectrum of Light

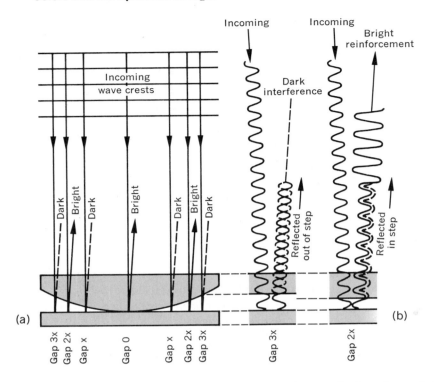

Fig. 12-7 (a) Waves of light (crests shown by lines) moving down on the two pieces of glass to form "Newton's rings." (b) The relation of the space between the two glasses to the location of successive bright and dark circles. The rays reflected from the upper glass are shown by solid lines; those reflected from the lower surface are dashed lines. Interference and reinforcement of light reflected from the upper glass and the lower glass depend on the wavelength of the light and the separation of the two glasses.

opposite effect takes place and they all speed up. The colors come out at the same angle, and we see white sunlight. You can easily draw what happens, according to the wave theory, in the prisms of Figure 12-2.

"Newton's rings" (Fig. 12-4) are formed when light shines on a thin, concave glass (like a watch crystal) set on a flat sheet of glass, as shown in cross section in Figure 12-7a. The gap between the two glasses is extremely thin, even at its widest part, but it can be measured. When white light shines on the two pieces of glass, concentric circles of different colors appear, violet near the center, green, yellow, and red farther out, then repeated rings. When light of only one color is used, rings of that color alternate with dark rings. With red light, each circle is larger than it is with violet light.

When red light falls on these two glasses, some of it is reflected off the watch glass, and some goes through and is reflected off the lower glass. Thus, from every point two rays are reflected up, as shown by the solid and dotted arrows in Figure 12-7b. A point on the first dark circle is shown at *A*, and *B* is a point on the first bright circle, seen from above. It was found that no matter what color of light is used, the space between the two glasses is twice as great at B as at A.

If the distance between the two glasses at A is x inches, then the ray reflected from the lower glass has to travel $2x$ inches farther than the ray reflected from the upper glass (x inches on the way in, and another x inches on the way out). We can explain the dark circle at A if the wavelength of the light we are using is $4x$. As shown in Figure 12-7c, at A the trough of one wave and the crest of the other travel to your eye together, because one wave is half a wavelength behind the other one. These two waves cancel each other. The circle of darkness is explained by the *interference* of the two waves.

At B, on the other hand, the space between the two glasses is $2x$ inches. The wave reflected from the lower surface has traveled $4x$ inches (a whole wavelength) farther than the ray reflected from the upper surface. The crests of both waves get to your eye together, making a crest twice as high. The bright ring marks where the two waves *reinforce* each other (Fig. 12-7d). Each successive dark ring is spaced where one ray travels $1/2$ a wavelength (or $3/2$, $5/2$, $7/2$, or any odd multiple of half a wavelength) farther than the other. Each bright ring lies where one reflected ray travels a multiple of a whole wavelength farther.

The wavelength of each color of light can be determined by measuring the gap between the two glasses at each point where a ring is formed. Measuring as best they could, both Newton and Huygens saw that the wavelength of red light would have to be about $3/100,000$ of an inch. Violet light would have even shorter waves; about 70,000 of them would fit into an inch. Huygens believed that it was perfectly possible for light to have such short wavelengths. Newton, on the other hand, felt that it was ridiculous to even consider that light could have such tiny wavelengths, and so he discarded this interpretation of rings. He clung to the particle theory of light, leaving the rings unexplained. Though Newton did not actively oppose the wave theory of light, it was obvious that he did not believe it. For over 125 years, his prestige and the great success of his laws of motion and gravitation kept most scientists from accepting the wave theory, or indeed, even taking it very seriously.

Then in 1801 an English physicist, Thomas Young, performed another experiment which could not be explained

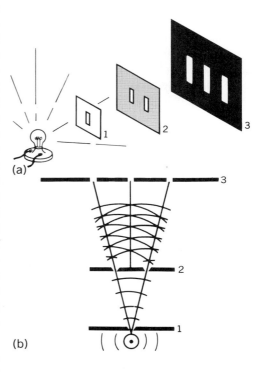

Fig. 12-8

Colors and the Spectrum of Light

by the particle theory, although its results were predicted by the wave theory. Young took a narrow beam of one color of light which had come through a narrow slit in a screen (Fig. 12-8a) and let it fall on a second screen which had two slits in it. Surprisingly, *three* slits of light appeared on the third screen beyond — one more in addition to the two you would expect. Between the two expected bright slits were two spots where the screen was completely dark, and midway between these dark patches was another bright slit. Between each bright patch and each dark one, the color gradually dimmed. If light were made up of small particles, like bullets shot from a light source, there could only be two bright slits, with darkness between them.

When water waves pass through a narrow opening, the opening acts like a new source of waves which spread out in circles from it, an effect called *diffraction*. If two boards with slits in them are placed in a tank of water and a stone is dropped back of the first board, a wave pattern results. Let us look more carefully at the new waves formed at each of the two slits (Fig. 12-8b). Notice that the waves from one of the slits cross those from the other slit. Along the three solid lines, crests of one set cross crests of the other, and troughs of one set cross troughs of the other. The waves *reinforce* each other along these three lines. Along the dotted lines, however, the waves cancel each other, or *interfere*, as crest meets trough and trough meets crest. The wave pattern from two slits is like that shown in Figure 12-9.

If we now consider Young's experiment, we see that reinforcement and interference of the two sets of waves would give a pattern like that in Figure 12-8b. There would be bright spots on the screen in the direction of the solid lines and darkness between them. When Young used light of a different color, the bright spots appeared at different places on the third screen, because the lines of reinforced and interfering waves were at different angles.

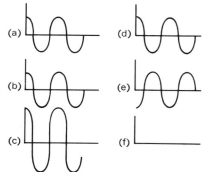

Fig. 12-9 As in Figs. 12-7 and 12-8, waves in phase ("in step") as in (a) and (b) add to give the larger wave in (c), illustrating *reinforcement*. The waves out of phase in (d) and (e) add to nothing (f), illustrating *interference*.

Knowing the distance between the two slits and the distance between the screens, he could calculate what wavelength could produce the bright spots at the observed positions on the third screen. His wavelengths agreed with those calculated from "Newton's rings."

When Young used white light, rather than light of one color, he got a spectrum of color like the rainbow in Figure 12-1. In fact, he got two spectra: one spectrum of violet to red on the right side of the central bright spot, and another going from blue to red in the other direction on the left side.

Figure 12-10 shows the spectrum of visible light. The wavelengths of light are so small that a special unit is used, the angstrom (A) which equals four-billionths

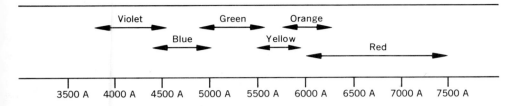

Fig. 12-10 The spectrum of visible light.

of an inch (0.4×10^{-8} in.). Reddish light extends from about 6200 to 7500 angstroms, and violet light from about 3800 A to 4500 A.

After Young's experiment, the particle theory of light was soon abandoned. Astronomers began to study the spectra of the light that came through their telescopes. Spectra can be obtained by prisms and photographed, or Young's method can be used, modified so that there are many slits instead of two. A sheet of glass, called a *diffraction grating*, ruled with many fine, parallel, closely-spaced scratches, can be used instead of a prism to get a spectrum. Light can pass through the glass only between the scratches. A shiny surface can also be ruled with many fine scratches, so that light is reflected only from the surface between the scratches. With more slits, you get more light of one color (waves of equal wavelength) going in one direction, and these can be focused by a lens. The combination of grating and lens gives several spectra in pairs, one on each side of the central white patch. Such spectra are brighter, because more sets of waves are reinforcing each other. One spectrum is picked for study, and the others (all similar) are blocked out.

If light is thought of as a wave motion, many things that it does can be explained, and the different wavelengths (colors) can be sorted out and studied. The wave theory works. But light travels through a vacuum as well as through transparent gases, liquids, and solids. In fact, it travels fastest through a vacuum, where there is *no* material. How do we reconcile this with Figure 12-5? What is waving? What corresponds to the water particles that move up and down to form water waves? Here the analogy with water waves breaks down. The best we can say is that we can understand most of what light does if we consider it to be wave motion.

Additional Reading

ABELL, G. O., *Exploration of the Universe*: New York, Holt, Rinehart and Winston, 1964, Chap. 7.

ASHFORD, T. A., *From Atoms to Stars*: New York, Holt, Rinehart and Winston, 1960, pp. 117-138.

PAGE, THORNTON, and L. W. PAGE, eds., *Starlight*: New York, The Macmillan Company, 1967.

chapter 13 | How Hot Is the Sun?

The beam of sunlight that fell on Newton's prism and the apple that fell in his garden were equally important. Newton's laws of motion and gravitation show how planets, satellites, and stars move, and where they are. As we shall see, studies of their spectra show what the conditions are like on their surfaces. Newton graciously said that because he "stood on the shoulders of giants" he could "see farther"—to his laws. (The giants were, of course, Ptolemy, Copernicus, Tycho, Kepler, and Galileo.) In the study of light, the situation was reversed: Newton's experiments with the sun's spectrum in 1666 first turned scientists' attention to it. Newton and others were the "giants" on whose shoulders later investigators stood. Their discoveries about light became the astronomer's means of finding out more about the universe.

In 1800, Sir William Herschel decided to continue Newton's experiments with prisms. Both heat and light come from the sun, and he wanted to find out whether they are spread equally through all parts of the sun's spectrum. He looked at coins and other objects through a microscope, lighting them first with one color of light from the sun's spectrum and then with another. He found that light from the yellowish-green part of the sun's spectrum was most intense; he could see the details of a coin much better with that light than when he used the other colors. As he went toward either the red or the violet ends of the spectrum, the coin was lit less and less and became harder and harder to see.

He placed a thermometer in the violet part of the spectrum for a few minutes, read it, and let it return to room temperature. Then he placed it in the green part, then in the yellow part, and so on, moving toward the red. The temperature of the thermometer rose a little bit faster as he tried it in each color nearer and nearer to red light. The thermometer rise in red light was 3½ times that in the violet light and 2¼ times that in the green light. He decided that in addition to the sun's *light* that he could see, there was *heat* in the sunlight.

Then he placed the thermometer beyond the red color, where there was no light visible on the table top. To his surprise, the temperature rose even more rapidly than it had in the red light. He moved the thermometer far beyond the red edge of the colored spectrum and it still warmed up. Here was a kind of sunlight, infrared radiation, invisible to your eye but felt by a thermometer and by your skin. You can feel the warmth radiating from a poker that has been in the fire, even though you do not actually touch the poker. The same sort of radiation from a radiator warms a room.

Because these invisible heat radiations are beyond the red end of the spectrum, they are called *infrared* (below the red). Herschel, who believed in the particle theory, saw that the infrared "particles" must be bent even less than those of red light as they enter and leave a prism. Today we describe infrared radiations as having longer wavelengths than those of red light. When a narrow part of the infrared spectrum was passed through two slits or gratings, cold and hot patches could be detected with a "dimming" of the heat in between. From the placement of these

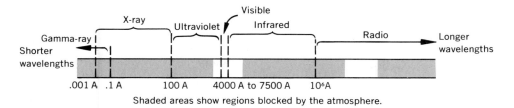

Shaded areas show regions blocked by the atmosphere.

Fig. 13-1 The spectrum of radiation (not drawn to scale) and the atmospheric "windows" (unshaded).

patches, the wavelengths were determined. The infrared region of the spectrum is a wide band of wavelengths extending from 7500 A to 10^5 or 10^6 A (Fig. 13-1).

Are there invisible radiations beyond the violet end of the spectrum too? Herschel asked himself this question, but was unable to answer it. However, at the same time he asked another question which evetually led to the answer of the first one: Is it possible that the *chemical* effects of the various colors of the spectrum differ, just as the effects on the eye (brightness of light) and on the skin (heat) differ? Johann Ritter, a German chemist, set out to answer this. He found that paper dipped in a solution of silver chloride was blackened by exposure to light — and thus photography was born. Ritter also found that red light blackened the paper less than green light. The blackening was greater (or took place more quickly) when he used blue light; and when he placed the silver chloride beyond the violet end of the spectrum, beyond visible light of the shortest wavelength, the blackening continued.

The wavelengths of this invisible, chemically active radiation proved to be from 4000 A to 3000 A, and the "light" was called *ultraviolet* (beyond the violet). Like infrared radiation, these rays can be "felt" but not seen. These are the wavelengths in sunlight that cause sunburn. Later, in laboratory experiments, ultraviolet radiation of even shorter wavelengths was produced, and by 1927 the ultraviolet range was extended to 100 A (Fig. 13-1). Waves shorter than 3000 A damage living things; therefore, they are used in hospitals to kill germs.

Since we're all alive, together with many other living things on earth, it isn't surprising that no ultraviolet light shorter than 3000 A gets to earth from the sun. It seemed likely, however, that the sun does produce such radiation, but that it cannot pass through the earth's atmosphere. Evidence to support this came in 1946 when a V-2 rocket launched by the U. S. Naval Research Laboratory carried a grating, or spectrograph, about 100 miles above the earth. It recorded ultraviolet wavelengths as short as 2200 A in the sun's spectrum. Since that time, radiation of even shorter wavelength has been recorded in spectra obtained in rockets and artificial satellites.

X-rays, of shorter wavelength than ultraviolet, were also produced in the laboratory, as well as the even shorter gamma rays, and have been recently identified in sunlight analyzed by spectrographs above the earth's atmosphere. Mercifully, we are shielded from these dangerous radiations by the air and special gases in the upper atmosphere.

How Hot Is the Sun?

Radio waves are also part of the spectrum. They extend from the end of the infrared portion; their wavelengths extend from 10^6 A (3.4×10^{-4} ft) to 1300 feet. Radio waves are the longest of all. The eye, the skin, and photographic film cannot detect them — nor can the ear. In the late 1880's, a German, Heinrich Hertz, first detected radio waves from electric sparks, and in 1906 they were used to carry telegraph messages (short and long bursts being used as "dots and dashes" in Morse code, just as in a telegraph signal carried by wires). It was first called the wireless. Then radio tubes were invented by an American, Lee De Forest, and these broadcasted continuous radio waves of one wavelength over an aerial. It is possible to "pulse" these waves very rapidly (hundreds or thousands of times per second) so that musical tones, human voices, and, since 1940, TV pictures can be transmitted by the pulses of radio waves from one antenna to another. Each radio station, or TV station, uses a different wavelength for broadcasting; and we tune in different broadcasts by selecting different wavelengths in the radio spectrum, from a few yards' length to as much as a mile.

In 1932 an American engineer, K. G. Jansky, built a radio antenna that picked up radio waves coming from one direction only. He found that when he turned this radio telescope toward the sky, away from man-made radio stations, he got "radio noise" from the stars. In 1942, the radio radiation from the sun was first received at radar stations in England.

All of these radiations, from long radio waves through the short waves of visible light and down to minute gamma-ray waves, travel at the same speed, 186,000 miles per second in a vacuum. Together they make up the spectrum of radiation shown in Figure 13-1. You can see that visible light is only a tiny portion of it. With our eyes alone, this is all that we would know of sunlight. Man's ingenuity in devising equipment and in making experiments has enabled him to detect and measure the rest of the spectrum of radiation coming from the sun and most other objects in the universe.

When you put a poker into a fire, the longer you keep it there, the hotter it gets, and the warmer your hand feels when you put it near the poker. The temperature of the poker has risen and the amount of infrared radiation that it is sending out has increased. If you heat it enough, a deep red glow becomes visible; shorter waves of red light are being radiated by the poker in sufficient quantity (intensity) to be seen.

The hotter the poker gets, the more energy it radiates. But how much more energy? In 1879, Josef Stefan, a professor at the University of Vienna, heated several solids and liquids to various temperatures and measured the energy radiated in all wavelengths added together. He discovered that there was a simple mathematical relation between the temperature and the total amount of energy radiated each second: $E \propto T^4$, when temperature T is measured on the "absolute" Kelvin scale.

In Stefan's time, as today, three temperature scales were in use. Most familiar to you is the Fahrenheit scale, on which water freezes at 32°F and boils at 212°F. Most scientists and almost everyone in Europe use the centigrade scale, where the freezing point of water is called 0°C and its boiling point 100°C.

Physicists had shown that the temperature of any object depends on the average motion of its molecules. The faster they are moving, the higher is the object's temperature. If the motion of the molecules were stopped entirely, the material should have zero temperature. It would be perfectly cold and could get no colder. They found by experiment that the pressure exerted by a gas on the walls of its container depends on the gas temperature. This pressure is caused by the impacts of gas molecules against the walls of the container. So if the temperature is lowered, the gas molecules go slower and the pressure becomes less. At lower speeds, each molecule hits the walls of the container less often and with smaller force.

Experiments showed (Fig. 13-2) that if a gas is cooled from $0°$ to $-1°C$, the pressure falls off by $\frac{1}{273}$. That is, if the pressure were 1 to start with, it would now be $\frac{272}{273}$ or 0.996. Each degree of cooling lowers the pressure by $\frac{1}{273}$. A $2°$ cooling produces a $\frac{2}{273}$ decrease in pressure (to 0.994); a $25°$ cooling produces a $\frac{25}{273}$ decrease in pressure (to 0.909); and so on. By $-150°C$ (pressure reduced by $\frac{150}{273}$ or to 0.450 of the original value), most gases condense to solids and the experiment has to stop. But by extrapolation you can see that if the gas were cooled by $273°$ (to $-273°C$), then the pressure would decrease by $\frac{273}{273}$ of its original value. (It would decrease to zero.) There would be *no* pressure, or in other words, *none* of the molecules would be hitting the side of the container. At $-273°C$ the molecules must be motionless; and if temperature measures the average motion of molecules, the gas has zero temperature.

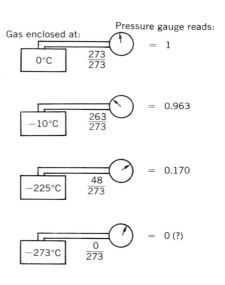

Fig. 13-2 The rate of decline of the pressure of a fixed amount of gas as it is cooled.

Yet the centigrade thermometer reads $-273°$ and the Fahrenheit thermometer reads $-459°$. So another temperature scale was set up, on which $-273°C = 0°$. This zero is called *absolute zero*, and the scale is called the *Kelvin temperature scale* (K). Each degree on the Kelvin scale is the same size as a degree on the centigrade scale, since the difference between freezing and boiling water is $100°$ on both. Fahrenheit degrees are smaller ($212° - 32° = 180°$ between freezing and boiling water). The Fahrenheit scale was based on the normal blood temperature in the human body, defined as $100°F$ (now known to be $98.6°F$) and on the coldest temperature Fahrenheit was able to obtain with a mixture of salt and crushed ice ($0°F$). But the Kelvin scale, starting at absolute zero, is based on something much more fundamental and hence is widely used in astronomy.

If we put the heated poker at the side of the fireplace, it cools. But it cools only to the temperature of the room. Then the long-wave radiation from other objects in the room brings energy to the poker at about the same rate that the poker radiates. Its molecules are still moving, although not as fast as when the poker was hot. If you could put the poker out in empty space so far from the sun or any star that no more heat came to it, then gradually all its heat would radiate away. The radiation would gradually become feebler and would finally dimish to nothing. Then the poker could grow no colder. It would have reached absolute zero temperature (0°K or −273°C or −459°F).

Thus it is not surprising that Stefan found a relationship between the rate of radiation in all wavelengths and the temperature on the Kelvin scale. He heated many materials to various temperatures and measured the radiation in all parts of their spectra. In every case, the energy rate was proportional to the fourth power of the Kelvin temperature reading: energy emitted per minute, $E \propto T^4$. Thus, if the temperature of the poker is doubled, say from 300°K to 600°K (that is, from 80°F to 621°F), the energy radiated per minute by the poker increases 16 times. As you have seen (pp. 72-73), the proportion sign $\propto$ can be replaced by an equals sign if we insert a constant k whose value has been determined. When the radiated energy is measured in calories per minute from each square centimeter of surface, $k = 76.8 \times 10^{-12}$. Stefan's law can then be written $E = 76.8 \times 10^{-12}T^4$ calories per square centimeter per minute.

Stefan couldn't measure every object that radiated energy. But since everything he could measure agreed, he extrapolated beyond experience and said that the relationship $E \propto T^4$ is a general law that applies to every radiating object. Later, when higher temperatures could be reached in the laboratory, the law still held. The temperature of red-hot iron is 920°K (1200°F) and the filament of an electric light bulb reaches 2770°K (4500°F), for instance. Increased energies radiated per minute are again found to be $76.8 \times 10^{-12}T^4$ for each square centimeter.

By extrapolating even further, we can measure the temperature of the sun, using Stefan's law. First of all, the total amount of energy that the earth receives from the sun can be measured: Each square centimeter of the earth (about ⅙ sq. in.) receives about two calories per minute when the sun is overhead. This is how much we receive, but how much does the sun send? Since the sun emits energy equally in all directions, we can imagine that this energy is spreading through a series of concentric spheres drawn around the sun.

The radiation that reaches the earth has spread out from a sphere of area $4\pi r^2$ (where r = the radius of the sun, 432,000 mi) to a sphere of radius $4\pi R^2$ (where R = the average distance of the earth from the sun, 93 million mi). The ratio of r to R is $^{432}/_{93,000}$, or $^1/_{215}$. This means that at the distance of the earth, the sun's radiation is spread over a sphere with area 215^2 (i.e., 46,225) times the area of the sun. Thus, the two calories per minute received by a square centimeter on earth correspond to 89,976 (i.e., $2 \times 46,225$) calories per minute from a square centimeter of the sun's surface. Using Stefan's law, this equals kT^4, and we can calculate T, the temperature of the sun's surface (as in Calculation 4) to be 5845°K, which is 5572°C or about 11,000°F.

Sir William Herschel had found that the greatest intensity in the sun's light is in the yellowish-green part of the spectrum. Later measurements showed it to be at wavelength 4750 A. Molten steel, at temperature about 1850°K, gives out the greatest amount of energy at much longer wavelengths (about 15,500 A infrared). The greatest amount of radiation from a glowing red-hot poker is in even longer infrared wavelengths at about 45,000 A. At cooler temperatures bodies not only radiate *less* energy, but they radiate most of it in *longer* wavelengths.

In 1896, William Wien, a German physicist, made accurate measurements of radiation from materials in his laboratory and found that the length of the waves of greatest intensity is inversely proportional to the Kelvin temperature: $\lambda_{max} \propto 1/T$. (The Greek letter lambda, λ, is commonly used by physicists to stand for wavelength.) Wien determined the value of the constant k that permits the equation to be written $\lambda_{max} = k/T$. It is 2.897×10^7 when the wavelength λ is in angstroms and the temperature is in degrees Kelvin. Now you can calculate the actual wavelength in which the most energy is radiated from bodies at various temperatures. For instance, you will find that a radiator with boiling water in it ($T = 100°C = 373°K$) radiates most at $\lambda_{max} = 77,500$ A, or about 0.0003 of an inch.

Wien, like Stefan, found that his law held to the highest temperatures he could reach in his laboratory, and it came to be considered a universal law. Astronomers began to use it as another method of measuring the sun's temperature. In the sun's spectrum, the wavelength of most intense radiation is at 4750 A. When this value is used in the equation $\lambda_{max} = k/T$, as in Calculation 5, the sun's temperature turns out to be 6100°K—pretty close to the temperature derived from Stefan's law.

Wien's law, $\lambda_{max} = k/T$, does not mean that the sun shines with yellow light only. As we have seen, its radiation contains all wavelengths of the spectrum—from short gamma rays to long radio waves. What it does mean is that the sun emits more yellow light than radiation of any other wavelength.

Stefan's law tells how much the total radiation will be at a given temperature. Wien's law tells which wavelength will have the greatest amount of radiation for each temperature. How is the remaining energy divided among all the other wavelengths of the spectrum? It can be measured, wavelength by wavelength, as shown in Figure 13-3, where the dots indicate the energy emitted at various wavelengths by a radiating object heated to a temperature of 1596°K. As you can see, these dots fall very nearly on a curve. The fact that they don't quite fall on a smooth curve was considered to be because neither wavelengths nor energy amounts could be measured precisely. Points on the smooth curve were considered to be the actual amounts of energy emitted at each wavelength.

In 1900, Max Planck, a German physicist, set out to find a mathematical for-

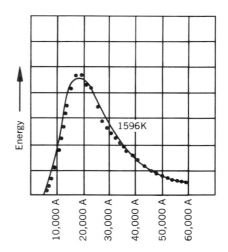

Fig. 13-3 The dots show the amount of energy (height above base of diagram) measured at various wavelengths (distance from left edge of diagram) from radiating material at 1596°K. The solid curve is drawn through the points predicted by Planck's formula.

How Hot Is the Sun?

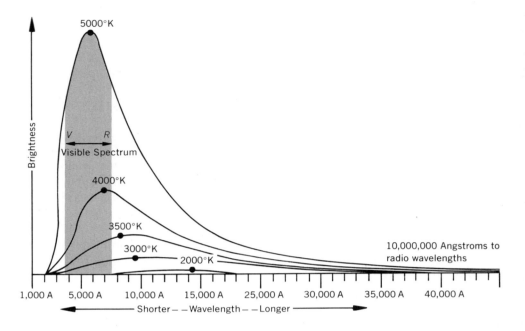

Fig. 13-4 Planck curves for various temperatures.

mula that would tell, for every temperature, where the "dot" should fall for each wavelength. In other words, he was seeking a formula that would predict every point on the curve for each temperature. This formula would be used to calculate every point on curves like those in Figure 13-4, which were drawn through points determined in laboratory experiments (Fig. 13-3). The formula turned out to be a complicated one, but it worked.

One of the great values of these curves is that they make determinations of the temperatures of celestial objects simpler and surer. It is difficult and time-consuming to measure the energy radiated in each wavelength and add them all up. Different equipment is needed to measure the energy of ultraviolet, visible light, and radio waves. Yet this must be done if Stefan's law is to be used. When Wien's law is used, the temperature determination rests on the accurate measurement of the wavelength where the curve peaks. Using the Planck curves, however, energy determinations for two or three measured wavelengths will tell on which Planck curve a spectrum fits. These two or three points will all fit on only one curve.

In an approximate way, we can see how this can be done. In Figure 13-4, the height of the 5000°K curve at wavelength 5500 A (yellowish-green) is 2½ inches. At 4000 A (violet light) the height of the curve is 1¾ inches. The ratio of these heights (proportional to their amounts of energy) is $2.50/1.75$, or 1.42. This is a measure of the color, and it can be used to determine the temperature of any star. The color index of the Planck curve for 4000°K is 0.25 in. / 1.00 in., or ¼. Therefore, instead of measuring the brightness at every wavelength and seeing which curve it

100

matches, the brightness at 4000 A and at 5500 A can be measured, their ratio calculated, and the particular Planck curve that this ratio fits can be found.

If you study the Planck curves in Figure 13-4, you will see that they show three things about sources of radiation: (1) A radiating body at any temperature gives out radiation at all wavelengths but not in equal amounts for each wavelength. (2) From one square centimeter (or other equal areas), a hotter radiating body sends out more radiation in every wavelength than does a cooler radiating body. (3) A hotter radiating body emits the largest proportion of its energy in shorter wavelengths than does a cooler one. Therefore, the color of a hotter body is different from that of a cooler one. Compare the hot poker glowing with red light at 920°K and a star, at 5000°K, whose light is slightly yellowish.

The light from the sun, analyzed by its spectrum from a prism or grating, is a long-range thermometer, the only kind astronomers can use to measure the temperature of the sun, 93 million miles away. In fact, distance makes no difference to this kind of thermometer. The temperature of many other shining objects in the universe can be measured with it.

Additional Reading

HUFFER, C. M., F. E. TRINKLEIN, and MARK BUNGE, *An Introduction to Astronomy:* New York, Holt, Rinehart and Winston, 1967, pp. 46-50, 168-170.

MOTZ, LLOYD, and ANNETA DUVEEN, *Essentials of Astronomy:* Belmont, California, Wadsworth Publishing Company, 1966, pp. 278-285.

chapter 14 | # The Temperatures of the Planets

Like the other planets, the earth gets all its light and heat from the sun, except for a very small warming effect due to radioactivity in the rocks below its surface. Measurements show that the subsolar point, the area directly under the sun (a place near the equator where it is noon), receives about two calories per square centimeter per minute.

The closer a planet is to the sun, the more of the sun's radiant energy reaches it. The amount is inversely proportional to the square of the planet's distance from the sun, $\propto 1/R^2$ (p. 60). The earth is 1 AU from the sun, and we know the other planets' relative distances (Table 5, col. 1). Therefore, we can calculate how much of the sun's radiant energy reaches the subsolar point on each. A square centimeter near that point on Mercury, at 0.39 AU from the sun, receives 13 calories per minute, or $2/(0.39)^2$. The amounts of solar energy which reach the other planets can be calculated in the same way and are shown in column 2 of Table 5.

	Distance from sun (in AU)	Calories received each minute by 1 sq cm near subsolar point	Percent of radiation retained	Calories retained each minute by 1 sq cm near subsolar point	Average temperature (K), as predicted by Stefan's Law	Measured temperatures (F)
Mercury	0.39	13.2	94	12.2	650° (+709°F)	+610° to −500°
Venus	0.72	3.85	24	0.925	234° (−38°F)	−27° to −45°
Earth	1.00	2.00	59	1.18	239° (−11°F)	40°
Mars	1.52	0.87	85	0.74	222° (−60°F)	80° to −150°
Jupiter	5.20	0.074	49	0.036	105° (−270°F)	−220°
Saturn	9.54	0.022	50	0.011	77° (−320°F)	−230°
Uranus	19	0.0055	33	0.0018	49° (−370°F)	below −300°
Neptune	30	0.0022	38	0.00083	41° (−385°F)	−350°
Pluto	39	0.00132	84?	0.0011	43° (−382°F)	below −350°

Table 5 Distances, heat received and absorbed, calculated and measured temperatures of the planets. (Measured temperatures listed for Venus refer to the top of the cloud layer; at the surface the temperature is about 640°F.)

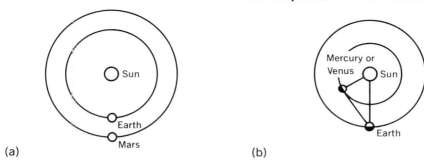

Fig. 14-1 (a) Mars at "opposition" is 180° from the sun and we see all of its lighted half to measure its albedo. (b) Mercury or Venus can be measured at quarter phase when we see half of the light reflected by the planet.

The planets shine by reflected sunlight, so we know that some of the solar energy they receive goes back out into space. We must estimate how much, in order to predict the amount of energy available to heat each planet. Astronomers can measure how much light reaches us from a planet and then compare it with the amount we would be getting if *all* the sunlight the planet receives were reflected back into space. The difference will be the amount that the planet absorbs.

We have already calculated the energy that each planet receives at its subsolar point. Now we must estimate how much the whole lighted half is receiving. It can be shown by geometry that the lighted half of a sphere (like a planet) receives the same amount of radiation as a circle with the same radius r would receive. The area of the circle πr^2 is only half the area of the curved hemisphere of the planet $2\pi r^2$; but every square centimeter of the circle receives the same radiation as the subsolar point, since it is a plane at right angles to the direction in which the sun's rays are traveling. Multiplying the circle's area (in square centimeters) by the number of calories received per minute at the subsolar point gives an accurate estimate of the energy received by the sunlit half of the planet.

If we choose a time when one of the outer planets (Mars, Jupiter, Saturn, Uranus, Neptune, or Pluto) is lined up with the earth and sun (Fig. 14-1a)—when the planet is 180° from the sun in the sky—we are viewing this whole lighted hemisphere. At that time, called *opposition,* the distance of the planet from us is just the difference between its distance from the sun and our distance from the sun. So if we measure the brightness of a planet at opposition and multiply that by the square of its distance from us, we get the amount of sunlight reflected from its whole lighted hemisphere. The fraction of sunlight reflected is called the *albedo,* usually given as a percent.

Of course, Mercury and Venus are closer to the sun than the earth is and can't be seen at opposition. They can be measured when we see exactly half of the sunlit hemisphere (Fig. 14-1b), and the distance from us can be calculated by geometry from the sun-planet distances. From the earth-planet distance, we can also calculate the true diameter of each planet after measuring its angular diameter (Fig. 14-2). The angular diameter of a planet changes as it comes closer or gets farther from the earth; the true diameters are, of course, always the same.

103

The Temperatures of the Planets

Fig. 14-2 The viewer on earth sees the planet with angular diameter $a°$. He knows the distance of the planet. In the long, thin triangle, he can solve for $2r$ (the diameter of the planet).

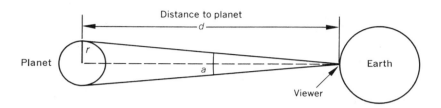

The albedo is always less than 100 percent. This means that all of the planets absorb some of the sun's energy and use it to heat themselves up above the absolute zero of outer space. The percentage of light absorbed by each planet is given in column 3 of Table 5. Notice the differences among the planets. We realize that these variations are caused by differences in the surfaces of the planets, recalling that snow or white sand reflects more light than does a dark rock or brown soil. It should not surprise you to learn that photographs of Venus, which reflects 76% of the sunlight falling on it, show it to be covered with dense clouds. You have seen clouds in our sky brilliantly reflecting the light of the setting sun. The dull, rocky surface of Mars (albedo 15%) visible beneath sparse dust clouds, is shown in recent space-probe photographs like Figure 14-3.

Knowing the number of calories received and the percentage absorbed by a planet, we can calculate how many calories are retained to heat the planet. Mercury, which receives 13 calories per square centimeter at its subsolar point, absorbs 94% of them. Thus, 12.2 calories per minute are available to heat that square centimeter: $(13)(0.94) = 12.2$. Table 5 (col. 4) gives this information for the other planets.

Now we can use Stefan's law, $E = kT^4$, to estimate the temperature of Mercury. The constant $k = 76.8 \times 10^{-12}$ (p. 98). Substituting the proper values in the equation, we can solve for T to get a temperature hot enough to melt tin and lead:

$$12.2 = (76.8 \times 10^{-12})\, T^4$$
$$T^4 = (12.2)/(76.8 \times 10^{-12}) = 1.59 \times 10^{11}$$
$$T = 631°K = 358°C = 676°F$$

The whole sunlit side of Mercury will not be this hot, of course, for much of the sunlight will be falling at an angle to the curved surface. But we would not expect it to vary by more than 100°C, and this still leaves the surface uncomfortably warm.

Mercury is rotating very slowly, about one turn every 59 days. It goes around the sun once in 88 days; and up until 1965, astronomers studying certain vague markings on the planet's surface thought that it also rotated once in 88 days.

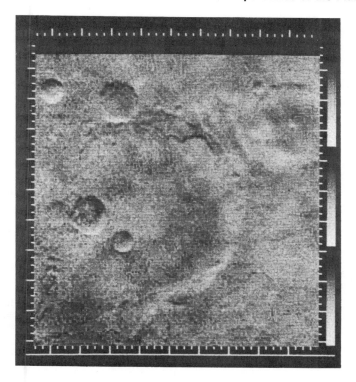

Fig. 14-3 A photograph of the surface of Mars made by Mariner IV on July 14, 1965 from 7800 miles above the planet, with the sun halfway up the Martian sky. The lines on the scale bordering the photograph are about 4.3 miles apart. [Courtesy Jet Propulsion Laboratory/National Aeronautics and Space Administration.]

Then one side of the planet would always face the sun, just as our moon always has the same side facing the earth (Fig. 14-4a). If Mercury rotated in 88 days, no solar energy would ever fall on the side which always faces away from the sun. That dark side would be warmed only by heat conducted from the lighted side through the planet and by radioactive processes going on within the planet. It would be very cold. However, the new measurements show that Mercury rotates faster than once in 88 days, so all sides get heated. Since the nights are very long—about 90 days (Fig. 14-4b)—the dark side gets almost as cold as it would if it were never lighted. If you ever went there, you would spend 90 days in the frying pan and the next 90 days in the deep freeze.

What temperatures does Stefan's law predict for the other planets? By observation of surface markings such as the "red spot" on Jupiter (Fig. 14-5) and many surface features of Mars, the rotation period or "day" on a planet can be determined (Table 6). All the planets except Mercury, Venus, and Pluto rotate fairly rapidly; and all longitudes on each receive sunlight for about half the time, from five hours at a stretch on Jupiter and Saturn up to about 12 hours on Mars.

105

The Temperatures of the Planets

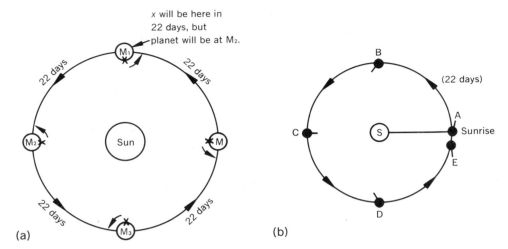

(a) (b)

Fig. 14-4 (a) Diagram of Mercury to illustrate how equal rotation and revolution would keep the same side of a planet facing the sun. (This also applies to the moon on its orbit around the earth.) (b) In 22 days Mercury travels 90° around its orbit (360° in 88 days), but a point on the planet, indicated by the spike, actually rotates through 135° in that time—about 3/8 of a full turn, since Mercury rotates 360° (a full turn) in 59 days. At A, it is just before sunrise at the spike; at B, 22 days later, it is midmorning; at C, it is noon; at D it is midafternoon; and at E (88 days later than A) the sun has just set.

As we have seen, the solar energy retained each moment to heat a planet is equal to that retained on a circle with area πr^2. However, this energy must heat the whole area of the planet's sphere, an area equal to $4\pi r^2$. Therefore, we must modify Stefan's equation to get the *average* temperature:

$$(\pi r^2)E = (kT_{av}^4)(4\pi r^2), \text{ or}$$
$$E = 4kT_{av}^4$$

Fig. 14-5 Jupiter showing the "red spot," 30,000 miles across (upper left), which has been observed since at least 1831 but is still not satisfactorily explained. The "stripes," light and dark bands parallel to the equator, show continuous but gradual changes, indicating that they are cloud bands rather than the solid surface of the planet. The high percentage of reflected light (51%) also indicates a cloud cover. [Yerkes Observatory photograph.]

106

	Period of revolution around the sun (= planet's year)	Period of rotation (= planet's day)
Mercury	88 days	59 days
Venus	247 days	Estimates vary from 22 hours to 30 days
Earth	365 days	24 hours
Mars	687 days	24½ hours
Jupiter	12 years	10 hours
Saturn	29.5 years	10 hours
Uranus	84 years	11 hours
Neptune	165 years	16 hours
Pluto	248 years	6 days

Table 6   Lengths of years and days on the planets, in terms of hours, days, and years on earth.

Thus, to estimate the average temperature of Jupiter, our calculations are:

$$0.036 = (4)(76.8 \times 10^{-12})(T^4)$$
$$T^4 = 0.036/(76.8 \times 10^{-12})(4) = 1.17 \times 10^8$$
$$T = 104°K = -169°C = -272°F$$

Column 5 of Table 5 shows the average temperatures of the planets predicted by Stefan's law. The actual temperatures are expected to vary above and below these values.

The sunlight absorbed by the planets heats them above absolute zero. Every object with a temperature above 0°K radiates energy. Therefore, the planets must each be producing radiation of their own. Can't we just measure the wavelength of maximum intensity in the spectrum of each planet and use Wien's law (p. 99) to find the planets' temperatures directly? Or can't we get the whole spectrum of a planet's radiation and see which Planck curve it fits?

The difficulty is that both the reflected sunlight and the planet's own radiation come to us together through the telescope. The two must be separated before the planet's radiation can be studied. The temperatures we have predicted range from 43°K for Pluto to 631°K for Mercury. Using Wien's law, $\lambda_{max} = k/T$, with $k = 2.897 \times 10^7$ (p. 99), we can find the wavelength of maximum intensity for Mercury: $\lambda_{max} = 2.897 \times 10^7/631 = 4.59 \times 10^4$ A. Pluto's wavelength of maximum intensity will be about 6.7×10^5 A, and those for the other planets will lie between these two extremes. Thus, the wavelength of maximum intensity of every planet is in the infrared (7500 A to 10^6 A). The Planck curves calculated for peak intensities like these (Fig. 14-6) show that no visible light is given out by the planets themselves. All of their radiation is in infrared and longer wavelengths.

The Temperatures of the Planets

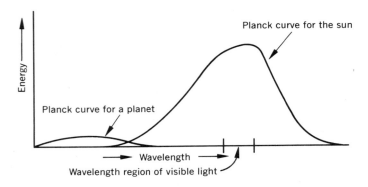

Fig. 14-6

An ingenious astronomer found that a water cell does not transmit long-wave radiation. By placing one in a telescope, the long-wave radiation of the planet can be separated from the shorter wave sunlight merely reflected from the planet's surface. The minor amount of long-wave radiation due to the reflected solar spectrum can be calculated and subtracted. Of course, the eye cannot see any wavelength longer than that of red light, but there are instruments called infrared detectors that can. They tell the astronomer how much radiation the planet is producing at each infrared wavelength and thus give a temperature reading for the planet, using Wien's law or Planck curves.

How well do these observed temperatures agree with those we have predicted? They are listed in column 6 of Table 5, and for all the planets except Saturn the agreement seems excellent. The temperatures determined for Mars are based on measurements of its infrared radiation taken at many places on the planet. Because of its relative closeness to the earth (at times it comes within 35 million mi), Mars has been photographed and studied more than any other planet. Its surface features (Fig. 14-7) are easily seen through even a fairly small telescope, indicating that its atmosphere is fairly thin and rather free of clouds. The polar caps, which appear to be mantles of frost around both its north and south poles, grow larger in the winter of each hemisphere. The greater part of the planet's surface is a rocky or sandy desert

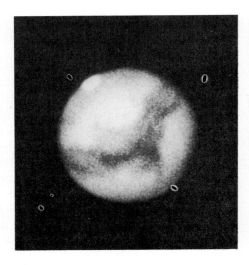

Fig. 14-7 Mars, showing polar cap and dark band around the equator. [Yerkes Observatory photograph.]

(pockmarked with the craters photographed by the Mariner IV space probe). Haze and yellow clouds cross the deserts from time to time. Around its equator is a dark, blue-gray belt that changes size with the Martian seasons.

As you would expect, the temperatures vary with the season, the location, and the time of day. The highest temperature recorded at the equator is about 80°F. During the Martian night, however, the temperature at the equator drops to nearly −100°F; at sunrise it is −4°F and at sunset 40°F. The north polar cap seldom gets above −100°F, while the south cap ranges from −150°F in winter to almost 32°F in summer. The temperature decreases gradually from the equator across the deserts to the poles.

So many determinations of Mars' temperature have been made that a contour map with lines drawn through points with equal temperature can be drawn for each Martian season. These maps resemble similar maps of earth's temperatures, found in many atlases. In addition, a few wind directions have been obtained from cloud-drift directions. When these are combined with the temperature maps, a rough "weather map" for Mars, resembling those put out by the U.S. Weather Bureau, can be drawn. They show the presence of a prevailing westerly wind in both hemispheres and hint at highs and lows like those on the earth.

Just as the sun emits radiation in all wavelengths even though the bulk of its radiation is in visible light, so do the planets send out radio waves as well as in-frared waves. The strength of the radio radiation from Venus has been measured and indicates a temperature of about 640°F! Because radio waves can penetrate the thick cloud layer of Venus, they probably come from the actual surface of the planet. This high surface temperature (in contrast to that of the upper atmo-sphere, −27°F to −45°F) was also measured by the Mariner II Venus probe in 1962. The thick clouds of Venus suggest a thick atmosphere that allows visible light to come in but prevents infrared radiation from escaping outward, like the water cell in the telescope. Earth's atmosphere has a similar warming effect on our planet (compare its predicted and measured average temperatures in Table 5). It is called the "greenhouse effect." Anyone who has sweltered in a green-house on a cold but sunny January day can testify to its efficiency The glass lets the light in, and the radiation heats the greenhouse and produces long-wave radia-tion that cannot escape through the glass so it, too, heats the greenhouse. Recent radio measurements deep below the clouds of Jupiter and Saturn show that the temperature increases with depth; it is possible that their surface temperatures may be much higher than their atmospheric temperatures.

Additional Reading

MOORE, PATRICK, *The Planets:* New York, The Macmillan Company, 1958.

PAGE, THORNTON, and L. W. PAGE, eds, *Neighbors of the Earth:* New York, The Macmillan Company, 1965, Chaps. 1-3.

| **The Planets' Atmospheres**

Mars and the earth seem to be the only planets with comfortable temperatures. More important, Mars seems to be the only other planet where the temperature range would allow *organic compounds* to form and endure. These complicated molecules, chiefly of carbon, hydrogen, nitrogen, and oxygen, make up all known living things, from lowly one-celled animals to man. Laboratory experiments show that these molecules are broken up, or do not even form, at high temperatures like those of Mercury and the surface of Venus (Table 5). They might, however, be present high in Venus' cooler atmosphere.

Organic molecules will not form unless the "building blocks" of organic compounds — molecules of carbon dioxide, methane, ammonia and certain others — are present in liquid or gaseous form, or unless there is water that can move them about. Otherwise, as solids, they cannot come together and combine to form the organic molecules. It is clear from Table 5 that there probably is no liquid water (freezing point 32°F) on the planets beyond Mars. In addition, laboratory experiments have shown that at the low temperatures of the outer planets, many of the building-block molecules will not be gases. We cannot expect them to be present in the atmospheres of these planets.

Living things need water to exist. In addition, all forms of animal life that we know need oxygen gas to survive, and plants use carbon dioxide gas. On the basis of temperature alone, these gases could be present in the atmospheres of Mars and Venus. But are they? Do the atmospheres of these planets contain life-giving gases like oxygen, carbon dioxide, and water? These are questions that must be settled before we consider whether there could be higher forms of life (even men) on Mars. On Venus, where life seems possible only high in the atmosphere, we can expect only the simpler forms of life.

Is there any basis, other than temperature, on which we can predict which gases could be present in the atmospheres of Mars or Venus? Can we corroborate that all these gases, friends and enemies of life, are there? Can we rule out the possibility that certain of them are present?

We have seen (p. 97) that certain laboratory experiments can be explained only if we consider that all gas molecules are in continual motion. The molecules that make up the atmospheres of Mars and the other planets must also be moving, darting every which way. As a molecule collides with another and rebounds, its direction of movement changes. At every moment, some molecules are moving upward, away from the planet's surface. Some of these, of course, will collide with other molecules and reverse their paths; others may move high up in the atmosphere. All the while a molecule is moving upward, it is being pulled back by the planet's gravity.

When you throw a stone up in the air, it always falls back to earth. The upward motion that you gave it is overcome by the downward acceleration of gravity. If you throw the stone harder so that it leaves your hand with a greater speed, it will rise farther. But the force of gravity soon slows down the stone, then stops it, and reverses its direction. What happens when, standing at sea level (where the downward acceleration of gravity is 32 ft per sec^2), you throw a ball upward with a speed of 64 feet per second? At the end of the first second it is moving upward at only 32 feet per second, for by that time the acceleration of gravity has reduced its speed by 32 feet per second. At the end of the second second the ball is motionless for an instant. At the end of the fourth second, the ball returns to the earth at a downward velocity of 64 feet per second.

How fast would you have to throw that ball, or shoot a bullet, or launch a space probe so that it never came down? How fast must an upward-moving molecule be going in order to escape from the earth's atmosphere?

The ball only rose 32 feet above the ground, increasing its distance from the center of the earth by the fraction 32/(4000)(5280), or 1.5×10^{-6}. Therefore, we weren't too far off in considering the acceleration of gravity, $a = GM/R^2$ (pp. 60-61), during the ball's entire round trip to be the same as that on the earth's surface. But at the much greater heights to which a molecule in the atmosphere would rise, R becomes much larger and the acceleration of gravity would be much less; it decreases gradually as the molecule moves upward. This decrease must be taken into account.

If a small mass m falls toward a planet of larger mass M, Newton's laws, $F = ma$ and $F = GMm/R^2$, tell us that m's acceleration a is equal to GM/R^2. (That is, $a = F/m = GMm/R^2m = GM/R^2$.) If m falls from an infinite distance, at first the value of R (m's distance from the center of the planet) is infinitely large, so that acceleration is infinitely small. As m falls, R gets smaller and smaller and ends up equal to r, the radius of the planet, when m lands on the planet's surface. Using the calculus invented by Newton, mathematicians can break up R into many small segments and determine the increase in speed over each segment. When they add them all up, they find that the speed v at r (when m reaches the planet's surface) is equal to $\sqrt{2GM/r}$, or $v^2 = 2GM/r$.

Cleverly the mathematicians then ran the movie film backward, so to speak, and showed that if you start m outward from the surface of a planet at a speed of $\sqrt{2GM/r}$, m would get out to infinity. Moving at this speed or greater, m would escape from the planet. Notice that m doesn't appear in the equation; the *velocity of escape* is the same for a molecule of hydrogen or ammonia, for a bullet or missile, or for a space probe. Notice, however, that while G (the force of attraction between two one-gram spheres 1 cm apart) is the same everywhere, M and r are different for each planet.

The masses of the planets have been calculated (Chap. 10) and their diameters have been determined (Fig. 14-2). Cavendish found the value of G (p. 73). If we substitute these values in the equation $v_{escape} = \sqrt{2GM/r}$, we can determine the escape velocities for the planets; they are shown in Table 7. Or we can first calculate earth's escape velocity, 7 miles per second. Those of the other planets will equal $7\sqrt{M/r}$, where M is the other planet's mass in terms of the earth's mass, and r is its radius in terms of the earth's radius, values which are listed in the first two columns of Table 7.

The Planets' Atmospheres

Planet	Mass	Radius	Escape velocity (mi per sec)
Mercury	0.05	0.37	2.5
Venus	0.81	0.96	6.5
Earth	1.00	1.00	7
Mars	0.11	0.52	3
Jupiter	317	10.9	37
Saturn	95.0	9.1	23
Uranus	14.5	3.7	13
Neptune	17.6	3.5	14
Pluto	1.?	0.5?	10?

Table 7 Masses and radii of the planets (as compared to earth) and velocities of escape.

Now we begin to see an explanation for the thin atmosphere of Mars (escape velocity 3 mi per sec) and the thick atmosphere of Jupiter (escape velocity 37 mi per sec). A gas molecule moving at 3 miles per second could escape from Mars; it would have to be moving over 12 times that fast to escape from Jupiter.

How fast can we expect the molecules in the planet's atmospheres to be moving? Will molecules of the same gas move as fast in Jupiter's cold atmosphere as they do in the warmer atmosphere of Mars? Will hydrogen molecules move as fast there as oxygen molecules do? Let us consider gas confined in a laboratory container (as is described on p. 97). Moving molecules strike the sides of the container and produce a pressure that can be measured. (Pressure is force per unit area.) Experiments performed before 1700 showed that the pressure increases equally with equal increases of temperature (in degrees K). The molecules must be moving faster so that each square centimeter gets hit more often, or they must be hitting the sides harder, or both, as the temperature rises.

An automobile driven at 40 miles per hour hits a telegraph pole harder than the same car does, moving at 10 miles per hour. It does not hit the pole four times harder, as you might expect, but 16 times harder. Laboratory experiments show that the energy of a moving object is proportional to the square of its velocity (Energy of motion $\propto v^2$). At the same time, there is a difference in the force with which an automobile and a bicycle, both moving at the same speed, hit you. Laboratory measurements have shown the difference is proportional to the relative mass of each (Energy of motion $\propto m$): A 10-pound stone hits you 10 times as hard as a 1-pound stone does, if both hit you at the same speed.

The energy of a moving body is thus equally dependent on m and on v^2, so we must write the proportion: Energy of motion $\propto mv^2$. (Notice that we are using the same line of reasoning that led Newton to consider gravity as proportional to the *product* of m_1 and m_2, *not* as the sum of $m_1 + m_2$: see p. 58) The constant that permits us to replace the $\propto$ sign with the $=$ sign has been shown by experiment to be $\frac{1}{2}$ (a pleasant change from the constants that we have been dealing with in the last few chapters). So, Energy of motion $= \frac{1}{2} mv^2$. Thus, two gases whose molecules are of different mass have different velocities at the same temperature, although their energy of motion is the same.

The formula $E = \frac{1}{2}mv^2$ can be written $E = mv^2/2$. Thus $v^2 = 2E/m$, and $v = \sqrt{2E/m}$. Two gases at the same temperature (where E is the same for both) have velocities inversely proportional to the square roots of their masses ($v \propto \sqrt{1/m}$). An oxygen molecule weighs 16 times as much as a hydrogen molecule; their velocities are in the ratio $1/\sqrt{16}$ to $1/\sqrt{1}$, or $\frac{1}{4}$ to $\frac{1}{1}$. The hydrogen molecule is moving four times as fast as the oxygen molecule at the same temperature. And as the temperature rises, both will move faster; as it lowers both will move more slowly, but they will keep these same relative velocities. The energy of motion is proportional to the gas temperature (above absolute zero).

Since chemists have determined the mass of various molecules as well as the number of molecules in any container and since they can measure the pressure that all the molecules produce on the sides of the container (due to the energy of motion), the equation $E = \frac{1}{2} mv^2$ can be solved for any kind of gas molecule at any temperature. At the temperature of the earth's atmosphere, about $250°K$, hydrogen molecules must have an average speed of 1.25 miles per second to produce the observed pressure. Oxygen molecules must average only a quarter of this speed, or 0.32 mile per second. Both of these average speeds are well below the earth's escape velocity of 7 miles per second.

However, not every molecule is moving at this average speed. The pressure produced by the millions of moving molecules at various temperatures only tells us how fast the *average* molecule of a particular gas is moving. Some are going faster, some slower. It can be shown by statistics that a measureable proportion of the molecules are always moving up to 10 times as fast as the average, and many others are moving much more slowly. If there ever were hydrogen molecules in the earth's atmosphere, at any moment only a small fraction of them were moving at speeds greater than 7 miles per second. But over a period of several million years — and there is evidence that our solar system has been in existence much longer than this — all earth's hydrogen would gradually have leaked away. Not so its oxygen, for the fast molecule speeds would be only 10×0.32, or 3.2 miles per second, well below the earth's escape velocity.

Figure 15-1a shows the speeds of the *fast molecules* of several gases (10 times the average speeds) at temperatures from $50°K$ to $1000°K$. In Figure 15-1b, on the same background, each planet is indicated by a dot, placed according to its temperature and escape velocity. Figure 15-1c puts this information together. A planet whose dot falls below the velocity line of a particular gas will have none of that gas in its atmosphere. If our theory is right, earth's atmosphere should contain no helium or hydrogen. But if earth's atmosphere were cooler, the dot would be farther to the left, and we would expect hydrogen to be present. If Mars' temperature were higher, for instance $600°K$, it could hold no oxygen. But at its observed temperature, about $300°K$, it should be able to hold oxygen.

The Planets' Atmospheres

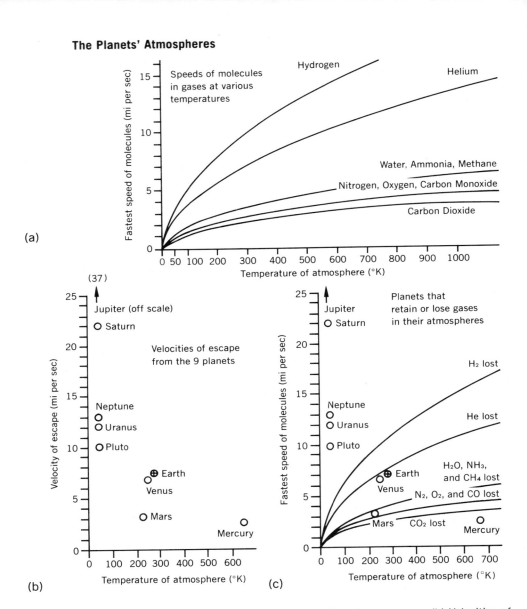

Fig. 15-1 (a) Speeds of the fast molecules in gases at various temperatures. (b) Velocities of escape from the nine planets. (c) The gases that will be retained or lost from the atmospheres of each of the nine planets.

Figure 15-1c predicts which gases can be present on the basis of each planet's escape velocity and the top speeds expected for each sort of molecule. However, at the low temperatures of the outer planets, some of these molecules (carbon dioxide and water in particular) would form solids rather than gases. Table 8 lists the gases that *could* be in the atmospheres of each planet, as predicted by freezing temperatures of the gases and the information in Figure 15-1c.

114

Carbon dioxide (CO_2):	Venus, Earth, Mars
Nitrogen (N_2):	All except Mercury
Oxygen (O_2):	All except Mercury
Carbon monoxide (CO):	All except Mercury
Water (H_2O):	Venus, Earth, Mars
Ammonia (NH_3):	All except Mercury and Mars
Methane (CH_4):	All except Mercury and Mars
Helium (He):	Only the outer planets
Hydrogen (H_2):	Only the outer planets

Table 8 Gases which could be in the atmosphere of the planets, as predicted by temperature and escape velocity.

Table 8 predicts the gases present in earth's atmosphere. However, analyses of our air show that it is made up of 78% nitrogen (N_2) and 21% oxygen (O_2). In addition there is 0.93% of a gas called argon. There are very small amounts of helium, hydrogen, carbon dioxide, and methane, and in certain parts of the world, exceedingly small amounts of ammonia gas and carbon monoxide. Together, all these make up only about 0.07% of dry air. There is water vapor present too, moving among the other molecules. Its amount varies from minute to minute and place to place, because when the relative humidity reaches 100% some of it forms cloud droplets. The low amount of helium is not surprising (Fig. 15-1c), but it is surprising that there is any hydrogen and that the other gases are so rare. Perhaps hydrogen gas is being added to the air constantly. Volcanoes or chemical changes in the rocks, or the breaking up of water molecules in the atmosphere could produce enough to offset the loss by escape velocity. Perhaps some of the other molecules, like ammonia and methane, were used up to form organic molecules in the early stages of evolution. These gases, deadly to higher forms of life, may be washed out of the air by rain water, or broken up by sunlight. In any case, they are just about missing from earth's atmosphere.

It seems likely that not all the gases we have predicted for the other planets are actually there either. Do Mars and Venus really have oxygen? Are the deadly carbon monoxide, methane, and ammonia gases there? By studying their spectra, astronomers can "take the planets' temperatures," even though the nearest thermometer is millions of miles away. With the nearest chemistry lab so far away, is there any way they can determine the composition of the other planets' atmospheres?

In the summer of 1801 a fire broke out in a looking glass factory in Munich, Germany. The flimsy building collapsed, killing all the workers except a 14-year-old apprentice named Joseph Fraunhofer. Seriously injured, he was rescued from the flames. The ruler of Bavaria happened to be passing by and took an interest in the boy, later presenting him with 18 ducats (about $40). Young Fraunhofer bought a machine to grind and polish glass into prisms and lenses, and over the next few years spent the remainder of the money for books on spectra and

The Planets' Atmospheres

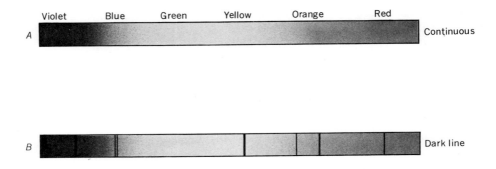

Fig. 15-2 (a) The continuous spectrum. (b) Dark-line, or absorption, spectrum.

light. Unwittingly, the Bavarian king had made a research grant whose dividends in scientific discovery, dollar for dollar, could well arouse envy in modern governments.

To Newton, the spectrum of the sun appeared as a continuous band of color (Fig. 15-2a). In 1802, an English scientist noticed several dark "lines" (Fig. 15-2b) running across the spectrum. He thought these were "the natural boundaries between colors." When Fraunhofer read this, he looked for these spectrum lines. Although they are difficult to see, by 1814 he had found about 600 of them in the sun's spectrum and had measured the exact positions in the spectrum (wavelengths) of 324 of them. Whatever instrument he used, these spectrum lines were always in the same place, forming a distinct, irregular pattern in the solar spectrum. He found that the spectra of the moon and planets also showed dark lines and that most of these were identical with those in the sun. Although he did not believe that these dark lines were color boundaries, he did not propose a theory to account for them.

Nevertheless, Fraunhofer's published descriptions and measurements led many scientists to look for these dark solar-spectrum lines. Among them was Gustav Kirchhoff, a professor at the University of Heidelberg in Germany. By 1859 he had found that dark spectral lines could be produced in the spectra of artificial light sources by passing their light through various gases. Each gas gave its own pattern of dark lines as individual as a fingerprint. He concluded that each gas removes certain wavelengths from the continuous white light spectrum, so that these wavelengths are missing or diminished, thus producing dark lines where colors would otherwise be. He called them *absorption lines* (Fig. 15-2b).

Today the wavelengths of many thousands of absorption lines in the solar spectrum have been measured on photographs where the lines show more clearly than when viewed by eye through a telescope. Each absorption line has been identified with lines produced in the laboratory by more than 60 chemical elements, alone or in chemical combinations with other elements.

The light by which we see each planet is reflected sunlight. If the planets were perfect mirrors and had no atmospheres, their spectra would be identical with that of the sun. But, as we have seen (Chap. 14), some of the sun's energy is absorbed and sent out to us as infrared and radio waves. In addition, the sunlight penetrates into each planet's atmosphere before it bounces back to us. In its passage into and out of these atmospheres, the light picks up more dark lines in its spectrum.

In the early 1940's a number of the dark lines in Jupiter's spectrum were identified. Light that was passed through a long pipe containing ammonia gas and through another containing methane gas gave the dark-line pattern observed in Jupiter's spectrum. Both of these gases were thus shown to be present in abundance in Jupiter's atmosphere. We can see absorption lines of ammonia in the spectra of that planet and Saturn, and also an indication of the methane in their atmospheres, and in the atmospheres of Uranus and Neptune. When compared to the sun's spectrum, those of Mars and Venus show the carbon dioxide in Venus' atmosphere and the lack of oxygen in the atmosphere of Mars. While it may be too early to conclude that there is no oxygen on Venus, nevertheless, all efforts to find it have failed. The gases so far identified by absorption lines in the spectra of the planets are listed in Table 9.

At the beginning of this chapter, Mars and Venus seemed the "best bets" among the other planets of the solar system as possible sites of life. Have the odds changed now that we are considering the composition of the planets' atmospheres as well as temperature? The outer planets, Jupiter, Saturn, Uranus, and Neptune, with their methane- and ammonia-laden atmospheres, look even less attractive. But what about Mars and Venus?

The lack of oxygen on Venus is significant. It is difficult to imagine how animal life could survive without it. Oxygen is very active chemically and is quickly removed from an atmosphere by chemical combination with other elements. Earth's atmosphere has a fairly large amount of oxygen only because plant life releases it continually, replacing the supply that is being used up by animals' breathing and other chemical reactions. The lack of oxygen makes it look as though there is no plant life on Venus.

What about Mars, legendary home of a sinister and skilled race of "little green men," racing through space aboard their "flying saucers"? Percival Lowell (p. 82),

	CO_2	N_2	O_2	CO	H_2O	NH_3	CH_4	H_2
Venus	✔	✔		✔	✔			
Mars	✔				✔			
Jupiter						✔	✔	✔
Saturn						✔	✔	
Uranus							✔	✔
Neptune							✔	✔

Table 9 Composition of the planets' atmospheres as shown by absorption lines. (No determinations have been made for Mercury or Pluto.)

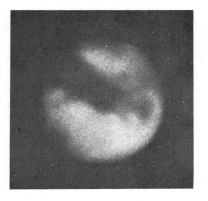 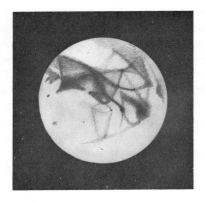

Fig. 15-3 Comparison of a drawing and a photograph of Mars made on the same night in 1926. Notice the "canals" in the drawing. [Lick Observatory photograph.]

the astronomer who laid the groundwork for the discovery of Pluto, believed that there might be intelligent life on Mars. Lowell and several other astronomers observed many fine, straight lines on Mars, which he believed were canals built by intelligent beings to carry melting ice from the polar caps (Fig. 14-7) across the desert to irrigate their crops. The peculiar thing is that although many reputable astronomers have since seen the canals clearly enough to draw them in the same positions, these lines have never shown up on any photograph of Mars (Fig. 15-3), even though a photograph usually shows more detail than does a glance through a telescope. The canals were not present on the photographs sent back by the Mariner IV space probe, and most astronomers agree that they are an optical illusion. The absence of oxygen on Mars makes it look as though the only possible life would be some primitive forms, such as lichens.

Additional Reading

> BRANLEY, F. M., *Mars: Planet Number Four:* New York, Crowell, 1962.
> SHKLOVSKII, I. S., and CARL SAGAN, *Intelligent Life in the Universe:* San Francisco, Calif., Holden-Day, Inc., 1966.

chapter 16 | **Shifts in the Spectra**

Back in the days when more people rode in trains, the whistle of a steam locomotive was a familiar sound. As a train sped by quiet farmhouses and country railroad crossings, people became accustomed to hearing the sound of the whistle change. As a train approached, the pitch of the whistle was higher than after the

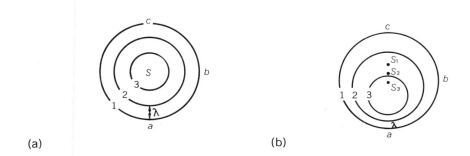

(a)   (b)

Fig. 16-1

train had passed. About the only time we can hear the same effect today is when an automobile passes with its horn blowing steadily.

By the early 1800's it was known that sound is a wave motion and that the pitch (or frequency) of a sound depends on its wavelength. It is inversely proportional to the wavelength ($\propto 1/\lambda$). In Figure 16-1a, sounds with wavelength λ are moving out from a source at S to listeners a, b, and c. The successive wave crests are labeled 1, 2, and 3. All three listeners hear sounds of the same pitch as several hundred crests like 1, 2, and 3 pass them. But it soon became apparent, at least to physicists, that if S were moving this would not be so.

In Figure 16-1b, S is moving with respect to listeners a, b, and c, and all the while is sending out sound of wavelength λ. When it was at S_1, it sent out the wave whose crestline is 1. However, S had moved toward a to S_2 when it sent out wave crest 2; to S_3 when it sent out wave crest 3. Thus, the wave crests passing listener a follow each other by a distance less than λ; to listener a the wavelength seems shorter than λ (and the pitch is higher than $1/\lambda$). The wave crests 1, 2, and 3 reach observer c with a distance between them greater than λ; to him the wavelength seems longer than λ (and the pitch is lower than $1/\lambda$). To listener b, at right angles to the path S_1 to S_2, the sound doesn't change pitch, because the distance between successive wave crests is still λ. Measurements show that the change in wavelength ($\Delta\lambda$) is described by the following equation:

$$\Delta\lambda = \lambda \ \frac{\text{velocity of the source}}{\text{velocity of the sound waves}}$$

In 1842, Christian Doppler, an Austrian physicist, pointed out that if light is a wave motion, there must be a similar lengthening or shortening of wavelength as a light source approaches or moves away, and the same equation should describe it: $\Delta\lambda = \lambda(v/c)$, where v is the velocity of the light source and c is the speed of light. For light *reflected* from a moving object, the fraction v/c would become $2v/c$. You could have guessed this from the fact that your image in a moving mirror seems to recede at $2v$ if the mirror is receding at v.

Shifts in the Spectra

Doppler searched for and found this change of wavelength. It is called the *Doppler shift* because each wavelength of light is shifted to another place in the spectrum. Doppler immediately saw its importance to astronomy. Not only does the wavelength shift tell us whether a light source, like a planet or a star, is moving relative to us, but it also tells whether it is moving toward or away from us. The speed of its movement in this direction can be determined: $v = c (\Delta\lambda / \lambda)$.

While the velocity of sound on a winter day is about 1100 feet per second (it varies with the temperature), the velocity of light is always 186,000 miles per second (9.8×10^8 ft per sec). This means that unless the light source is moving very fast, the fraction v/c will be extremely small. And the smaller this fraction is, the smaller $\Delta\lambda$ will be, since $\Delta\lambda = \lambda (v/c)$. We can expect v/c to be very small for movements in the solar system. Mercury, the swiftest planet, moves in its orbit at an average speed of about 30 miles per second, meanwhile reflecting sunlight to us from its surface. Thus, $2v/c = 60/186,000 = 3.2 \times 10^{-4}$. This value will be less if Mercury is not moving straight toward us. Also, if the earth is moving in its orbit toward Mercury, some fraction of our orbital speed (about 19 mi per sec) must be subtracted from $2v$. (The earth's rotation can be ignored if Mercury is observed near the meridian—see page 17—for then our rotation is carrying the telescope sideways only.) Because the earth and Mercury circle the sun in the same direction (as do all the planets), the earth's orbital velocity always reduces Mercury's Doppler shift to less than $2v/c$, whether it be positive (red-shift of recession) or negative (blue-shift of approach). The same is true for Mars and the other outside planets, where the Doppler shift is mostly due to the earth's orbital motion, $\Delta\lambda = \lambda v_E/c = \lambda (19/186,000) = \lambda (1.02 \times 10^{-4})$.

The wavelength of red light is about 7000 angstroms. The greatest shift ($\Delta\lambda$) that Mercury's relative motion could cause is 2.8 A. That is, $\Delta\lambda = (7 \times 10^3)(4 \times 10^{-4}) = 2.8$ A. The shorter wavelengths toward the blue end of the spectrum of visible light ($\lambda = 4000$ A) would be shifted by about half as much. The entire Planck curve for sunlight reflected from Mercury will show tiny shifts like these. The curve will move so slightly to the left or to the right that the color of the sunlight will not be changed. Only if v were in tens of thousands of miles per second would the curve be shifted enough to make the light appear redder or bluer than normal.

Shifts due to solar-system movements would be almost impossible to detect on a continuous spectrum. But fortunately, by Doppler's day, Fraunhofer had shown that the dark spectral absorption lines (p. 116) are real. Shifts in these lines are much easier to see and measure. Molecules of oxygen in the earth's atmosphere leave their "fingerprint," prominent dark lines, on the spectra of light received firsthand from the sun and on the second-hand sunlight reflected from Mars as it

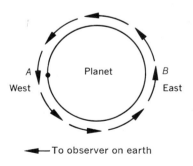

Fig. 16-2

approaches or recedes from earth. Oxygen molecules in Mars' atmosphere would leave this same pattern, too. However, each O_2 absorption line would be shifted slightly because of the relative motion of Mars and the earth. When Mars and the earth were getting closer to each other, this shift would show up as a thickening of the O_2 lines on their blueward sides. When Mars and the earth were becoming more distant, oxygen in Mars' atmosphere would produce a similar thickening on the redward side of each line. Because spectra show no such thickening, we can conclude that Mars' atmosphere has very little oxygen, if any.

We can also use Doppler shifts to indicate the orbital speeds of different bodies. The shifts in the spectra of Mars approaching and receding from earth (measurable in angstroms on the spectra) indicate velocities in the line of sight of 8.5 miles per second toward and 7.7 miles per second away, respectively. Orbital speeds, determined from Doppler shifts, are in good agreement with those predicted by Newton's laws.

Because the planets rotate, any point on their surfaces is sometimes moving toward us and sometimes away from us. Mars' radius r is 2000 miles and its rotation period P is 24½ hours. The velocity v of a point on Mars' equator at A in Figure 16-2 and given by the formula $2\pi r/P$, is 500 miles per hour or 0.14 mile per second toward us. At B it is 0.14 mile per second away from us. The ratio $2v/c$ is $^{0.28}/_{186,000}$, or 1.5×10^{-6}. This produces a blue-shift for A and a red-shift for B. The resulting shifts in wavelength (about 10^{-2} angstroms toward blue or red) for dark lines in the infrared portion of the spectrum (where wavelengths are about 7000 Å) are extremely difficult to measure accurately. About the greatest enlargement that a spectral photograph can stand, without becoming too fuzzy to measure, gives eight angstroms to the inch. The lines would be shifted $10^{-2}/8$, or 0.00125, inch toward the blue on the east side of Mars (approaching earth) and toward the red on the west (receding) side.

The rotation period of Mars can be determined much more easily and reliably by timing the movement of surface markings (Fig. 14-7) across the planet's disk, day after day. Astronomers do not have to rely on Doppler-shift measurements. Not so for cloud-shrouded Venus. Until 1965 astronomers could only guess at her rotation period. No trace of a rotational Doppler shift of the dark lines in Venus' spectrum had ever been observed. This meant that Venus is turning very slowly. Estimates of how small a shift could be detected indicated that one full turn of Venus could not take less than two weeks.

Then a radar beam of known frequency was directed to point A (Fig. 16-2) on Venus. *Frequency* is the number of waves per second; it is inversely proportional to the wavelength ($\propto 1/\lambda$) for all waves which, like light and radio, are part of the spectrum of radiation (Fig. 13-1) and travel at 186,000 miles per second. Radar frequency can be timed much more accurately than spectral wavelength can be measured.

The radar beam was found to bounce back from the edges of Venus with a measureably different frequency. After corrections for orbital movements, it indicated a Doppler shift at radar wavelengths (around 4 in.) of 2.4×10^{-9} inches. This, in turn, indicates a rotational speed at A (Fig. 16-2) of only 7 inches per

121

second. This speed shows Venus' rotation period to be 247 days. Venus' day is longer than her year (224 days)! Even more surprising is that point *A* was retreating rather than advancing. This means that Venus rotates from east to west, opposite to all the other planets and opposite to her own (and their) orbital motion.

Similar radar measurements have recently shown that a day on Mercury is 59 earth days. The long-held opinion that Mercury's day and year are both 88 of our days in length was based on the time intervals between the appearance of a few vague markings on the planet's disk. It was known that these were not permanent markings, but they were thought to have survived a few rotations. Apparently, in each case, new markings were misidentified with older ones which had disappeared.

When Galileo saw the planet Saturn, he turned from his telescope and said, "It is threefold." It did not appear as a round disk like the other planets, but like a main planet with companion planets on either side. It must have looked to him very much as it does in Figure 16-3, a modern telescopic photograph in which Saturn is overexposed in order to bring out four of the planet's moons.

Not until 157 years later, in 1766, did the telescope of the Dutch astronomer, Huygens (p. 87), reveal Saturn in all its glory, surrounded by gleaming rings (Fig. 16-4). The rings extend about 100,000 miles beyond the planet and are separated from it by a 7000-mile gap, yet they are only 25 miles thick. Doppler shifts (indicated by the arrows) show the planet itself to be rotating with a period of about 10 hours, and tell us about the nature of the rings. The outer edge of the

Fig. 16-3 Saturn (overexposed) and four of its satellites, from top to bottom, Titan, Rhea, Dione, and Tethys. The photograph was taken with the 82-inch reflecting telescope of the McDonald Observatory. [Courtesy Yerkes Observatory.]

rings appears to be rotating with a period of almost 15 hours and the inner edge with a period of about 5 hours. If the rings were a solid, *rotating* sheet of material, their period would be the same throughout (just as a planet has one definite period of rotation), and the outer part would have to be moving at a higher speed than the inner part. Instead, the speed of the outer part of the rings is less than that of the inner part. This shows that the rings are composed of millions of separate particles, and each particle is *revolving* around the center of the planet in accordance with Kepler's third law, $P^2 \propto R^3$ (p. 48), just like a tiny moon.

Distant stars can be seen through the rings, and this shows that there are gaps between the particles. Also, the rings contain no gas, since the spectrum of sunlight reflected from them shows no dark lines, except those due to the atmospheres of the sun and the earth. It seems evident that they are widely spaced, solid par-

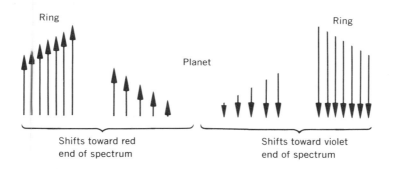

Fig. 16-4 Saturn and its ring system. [Mount Wilson and Palomar Observatories photograph.]

ticles, perhaps of snow, gravel, or sand, each in orbit around the center of the planet. Sunlight reflected from these particles of "earthy" material, commonplace enough to satisfy Galileo, produces golden, shimmering rings, celestial enough to satisfy Aristotle.

Additional Reading

ABELL, G. O., *Exploration of the Universe:* New York, Holt, Rinehart and Winston, 1964, pp. 159-161 and Chap. 12.

WHIPPLE, F. L., *Earth, Moon, and Planets:* Cambridge, Mass., Harvard University Press, 1963.

123

chapter 17

Small Planets, Meteors, and Comets

In 1766, when Saturn was the most distant planet yet discovered, a little known German, Johann Titius, somehow hit upon a scheme for remembering the distances of the planets from the sun. He took the numbers 0, 3, 6, 12, and so on (each number after the 0 being double the preceding one), added 4 to each, and divided by 10. The result is shown in the first seven lines of Table 10. Six years later, Johann Bode (p. 77), Director of the Berlin Observatory, heard of this. He was immediately struck by the blank at 2.8 AU and organized 24 European astronomers into a sort of committee to hunt for the missing planet.

As you know, Uranus was discovered in 1781, but it lay 19.2 AU from the sun. It was not the planet that Bode's men were looking for. However, it fit in well with "Bode's law," as Titius' scheme is known today, and stimulated the search for a planet at 2.8 AU. The discovery of Neptune, a maverick from Bode's law (as is Pluto) might have dissolved the committee, but by that time the gap at 2.8 AU was filled.

A Sicilian astronomer named Giuseppe Piazzi spent New Year's night in 1801 locating and mapping each star in the constellation Taurus. He was using a newly published catalogue in which each star was listed according to its latitude and longitude on the celestial sphere. Searching carefully for a listed star that he could not find in the sky (later shown to be a printer's error), his attention was caught by a faint "star" which the catalogue failed to list.

To his surprise, this "star" was in a slightly different position the next night. He looked at it every clear night and saw that it was moving eastward. Then in early February it reversed its direction, "looping" like a planet (p. 24). In November,

Titus' progression (Bode's law)	Planet	Planet's actual distance (in AU)
(0 + 4)/10 = 0.4	Mercury	0.387
(3 + 4)/10 = 0.7	Venus	0.723
(6 + 4)/10 = 1	Earth	1.000
(12 + 4)/10 = 1.6	Mars	1.524
(24 + 4)/10 = 2.8	———	———
(48 + 4)/10 = 5.2	Jupiter	5.203
(96 + 4)/10 = 10.0	Saturn	9.539
(192 + 4)/10 = 19.6	Uranus	19.191
(384 + 4)/10 = 38.8	Neptune	30.071
(768 + 4)/10 = 77.2	Pluto	39.518

Table 10 Bode's Law.

when the calculations of this "wanderer's" orbit were completed, it was found to lie at just about 2.8 AU from the sun. It was named Ceres (for the Roman goddess of Sicily) and joyfully hailed as the missing planet. Measurements of its angular diameter showed that Ceres is only about 500 miles across—a disappointingly small size, This should not surprise you; for although its orbit lay between those of Mars and Jupiter, both easily visible to the naked eye, it took a careful search with a good telescope to reveal Ceres.

Then, quite unexpectedly, another planet even more minor (300 miles across) was discovered a little over a year later. It also circles the sun at about 2.8 AU. By 1807 two more had been found, one with a diameter of 250 miles and the other with a diameter of 120 miles. By 1890 more than 300, all smaller than the first four, had been added to the list of minor planets. Today these are sometimes called *asteroids*.

Then the list grew rapidly after photography made discovery easier. When a long time-exposure photograph is made with a telescope, a motor moves the telescope tube westward, following the stars as they move across the sky during the night. This makes the stars appear as points of light (not as trails as they do in Fig. 2-2a, b, where the motor "drive" of the telescope was not used). However, even with the telescope drive working, a minor planet moving in its orbit makes a short trail, like the blurred image of a person in a group photograph who moved while the picture was being taken. Thousands of these short streaks made by minor planets have been seen on long-exposure photographs of the night sky near the ecliptic. From these, the orbits of about 3000 minor planets have been calculated. All of them are smaller bodies than the first four discovered, and most of them circle the sun at distances between 2.3 and 3.3 AU. There may be about 40,000 minor planets waiting to be photographed with powerful telescopes and to have their orbits calculated.

The sand-and-gravel-sized particles in Saturn's rings (Fig. 16-4) revolve around the center of that planet, each in its own orbit. In a similar way, each of these larger fragments, the minor planets, revolves in its own orbit around the sun (Fig. 17-1). Most of the orbits lie nearly in the plane of the earth's orbit, and their average eccentricity is not much greater than those of the planets.

Only the angular diameters of the four largest minor planets are big enough to measure, and even then it is very difficult. The angular diameter of Ceres (500 mi across and 180 million mi away) is only 0.005". In order to find the diameter (in miles) of the others, astronomers had to use a line of reasoning that should be

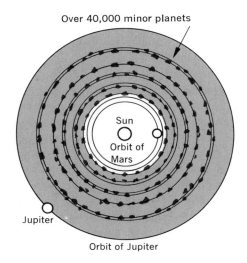

Over 40,000 minor planets

Sun

Orbit of Mars

Jupiter

Orbit of Jupiter

Fig. 17-1 Region of the minor planets.

familiar to you from the opening pages of Chapter 14. They knew how much sunlight reaches Ceres because they had determined its distance and size. They could measure how much reflected sunlight we, at a known distance from Ceres, receive. This, in turn, told them how much sunlight Ceres reflected: 10% of the amount that falls on it. The reflecting power of the three other large minor planets turned out to be the same. Since it is the same for the "big four" of this minor league, it seemed reasonable to conclude that it is the same for the others. Their distances from the sun and from us were known, so it was possible to calculate how big each minor planet must be in order to look as bright as it does. A minor planet, at the same distance from us as Ceres, which looks $1/100$ as bright must have $1/100$ Ceres' area, or about $1/10$ Ceres' diameter. One that is twice as far from us and looks $1/100$ as bright has $1/25$ the area or $1/5$ the diameter of Ceres, because distance dims the light by $1/r^2$.

Fig. 17-2 A meteor streaks across the sky. [Yerkes Observatory photograph.]

There are only a dozen minor planets with diameters of 100 miles or more and a few hundred between 25 and 100 miles across. The vast majority have diameters of a mile or so. If each cubic centimeter has about the same mass as a cubic centimeter of the moon, the combined mass of all known minor planets is only about $1/1600$ that of the earth. Of course, there is no reason to suppose that there are no minor planets less than a mile across since we cannot see those that are smaller and so far away. But even then, it is hard to see how their combined masses can be more than $1/1000$ that of the earth.

Many astronomers believe that the minor planets formed from the breakup of one planet — the one Bode's committee was looking for. The irregular shapes of the minor planets do suggest that they are explosion fragments, or that they have been nicked by collisions, or both. As a spherical planet rotates, the amount of light that we receive from it remains the same. But the light of many minor planets varies slightly, indicating that surfaces of different area are turned toward us as the minor planet rotates. An irregular, brick-shaped body would give the light variations measured for many of the minor planets. If the minor planets did at one time make up one planet, it was a very small planet indeed, unless some of them have gotten away.

And some of them have gotten away — those in the four gaps pictured in Figure 17-1. A small percentage of these escapees have collided with the earth. What

Fig. 17-3 Air photograph of a meteor crater near Winslow, Arizona. [Courtesy Yerkes Observatory.]

happens when the minor planet collides with the earth? We see a bright "shooting star" or meteor streak across the sky (Fig. 17-2). It might land as a meteorite and eventually be placed in a museum. It might dig a crater (Figs. 14-3 and 17-3) as it lands.

About 25 meteorites are found each year. Some of them are newly fallen; others are merely newly discovered. Most of them are stony, somewhat like the rocks of the earth's crust. Others are metallic, composed chiefly of iron and nickel, and some rare ones combine both types. They are the only objects from the universe beyond the earth that man has yet held in his hand or analyzed in the laboratory. Even if the meteor does not get picked up, its spectrum can be photographed in the brief moment it lights up the sky.

Before a meteor enters the air, its temperature is about the same as that of the moon's sunlit side (about 250°K); and it shines dimly by reflected sunlight until it enters the earth's shadow, where it does not shine at all. However, while meteors are plunging through the earth's atmosphere, they give out their own light (Fig. 17-4). The glossy surfaces of meteorites show that their outsides have been melted by friction with air molecules. Studies of meteor spectra show that part of the surface material becomes gaseous. The nearby air molecules are heated

127

too. Mixed with glowing gaseous material ripped from the meteor, they form, for a few moments, a brilliant trail of light behind the swiftly moving meteor. Of course, many small ones "burn up" long before they reach the earth's surface.

Spectra of meteors and their trails show bright lines rather than the dark lines with which you have become familiar. Kirchhoff (p. 116) found that spectra like this are made by glowing gases at high temperatures. The bright lines have the same wavelengths as the absorption (dark-line) spectra of the same gas. He called them *emission lines.* Thus, there are three sorts of spectra: continuous spectra given out by hot solids and liquids, absorption (dark-line) spectra characteristic of cool gases, and emission (bright-line) spectra indicating gases at high temperatures. Because the dark lines and the bright lines of a certain gas al-

Fig. 17-4 Meteor shower photographed from Kitt Peak by David McLean. [Sky and Telescope Magazine.]

ways appear at the same places in the same places in the spectrum (have the same wavelengths), Kirchhoff concluded that each particular gas can absorb or give out only certain wavelengths of light, thus producing dark lines or bright ones.

Comparison with laboratory spectra show that the emission lines of meteors and their trails are chiefly those of iron, calcium, silicon, aluminum, and sodium in gaseous form—the same elements shown by chemical analyses to be present in meteorites. They indicate temperatures of thousands of degrees. The bright lines of nitrogen and hydrogen are made by gases in the earth's atmosphere, heated to the glowing point.

Radar Doppler shifts and the motion across the sky give the speed and direction of a meteor's motion. From this information, astronomers can calculate the orbit that the meteor had been following before being pulled in by the earth's gravity. Many had orbits like those expected of a perturbed minor planet. The arrival of these meteors cannot be predicted. They usually occur singly; at the most, four or five can be observed from one place in a single night, each with a *slightly* different orbit. Over the whole earth, many thousands can be seen each night.

Several times each year, however, on certain definite dates, there is a night or two when a great number of meteors come in, all from the same direction. Meteors from these *showers,* as they are called, never land as meteorites. They are so small that they burn up completely in the atmosphere. On these dates, it is as if the earth were crossing a crowded boulevard where it is smacked by a heavy cross traffic of light cars, rather than a turnpike with a few big trucks that may or

Date of best display	Name	Associated comet	Period of comet (yrs)
January 3	Quadrantid	————	7.0
April 21	Lyrid	1861-I	415
May 4	Eta Aquarid	Halley	76
July 30	Delta Aquarid	————	3.6
August 11	Perseid	1862-III	105
October 9	Draconid	Giacobini-Zinner	6.6
October 20	Orionid	Halley	76
October 31	Taurid	Encke	3.3
November 14	Andromedid	Biela	6.6
November 16	Leonid	1866-I	33
December 13	Geminid	————	1.6

Table 11 Dates of meteor showers.

may not reach the intersection at the same time as the earth. Put another way, earth's orbit crosses an orbit used by not just one fragment, but by a long, thin swarm of many fragments, so that whenever we cross it there are meteors aplenty.

The orbits of "shower" meteors around the sun are extremely long ellipses, with the sun very much off center. In 1866 it was discovered that the meteors that shower down on us each year on August 11 travel in the same orbit as Comet 1862-II. Since then, almost every meteor-shower orbit has been found to be identical with an earlier known comet's orbit, as shown in Table 11. Twice each year, the earth passes through the orbit of the most famous comet of all, Halley's comet, providing the May and October showers. Most comet orbits are highly tilted to the earth's orbit, however, so that we can cross them only once each year, and many come nowhere near the earth's orbit.

The most important traveler along such an orbit is the comet (Fig. 17-5). Spread out along the rest of the orbit are fragments that can be no larger than gravel or sand grains because they do not survive the trip through our atmosphere. Their bright-line spectra, when they shine as meteors, indicate that they are rocky material. It is as if the comet, moving along its orbit, were falling apart, with its rocky debris trailing behind it all along the orbit, like papers in the wind behind a car with a litterbug in it. No matter where we cross the orbit, there is some debris ready to be captured by the earth's gravity and fall into the atmosphere as meteors.

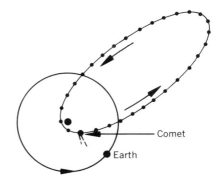

Fig. 17-5

Small Planets, Meteors, and Comets

A meteor, whether of comet or minor-planet origin, glows as it is heated by its speed through the earth's atmosphere. The light of a comet, however, is entirely unconnected with the earth's atmosphere. It is caused by radiant energy from the sun. Far from the sun, where a comet spends most of its time, all of its material is extremely cold. As it comes a bit closer to the sun we can see it faintly because it reflects sunlight. (Most comets are then less than 200 miles in diameter.) Its spectrum is that of sunlight reflected from a solid gray material. From the debris that falls as meteors, it looks as though the comet is a swarm of fragments, rather than one big chunk. But how could the swarm stay together?

When a comet has moved to within a few AU of the sun, its material begins to evaporate and these gases begin to glow, forming a cloud of light around the comet that may reach a hundred thousand miles in diameter. The spectrum now shows bright lines that indicate the presence of carbon, hydrogen, nitrogen, and oxygen, mostly in the form of molecules. Fred L. Whipple, Director of the Smithsonian Astrophysical Observatory, pointed out in 1952 that these elements make up methane (CH_4), ammonia (NH_3), and water, all of which would have been frozen solid farther out in space. In a comet, they would form a single ball of ice and wax that would hold the rock fragments together.

When a comet is only a few hundred thousand miles from the sun, where the temperature is about 4500°K, the comet's spectrum begins to show the bright lines of glowing sodium, iron, silicon, and magnesium atoms, elements which make up rocks. This gives additional evidence that some of the solid chunks are rocks. Most comets develop tails (Fig. 11-3d) as they approach the sun. The tail consists of the glowing gases detected by its spectrum and may be more than a hundred million miles long.

Whipple's theory, now generally accepted, is that a comet is like a dirty snowball, one in which pebbles and sand are mixed with the snow. When you bring the snowball in and set it near the fireplace, the snow evaporates and leaves a small pile of gravel and sand on the hearth; but comets do not stay near the sun long enough to melt completely.

We see a comet far off, glowing in the sky night after night, slowly moving with respect to the stars as it travels its orbit around the sun. In time, its orbital motion carries it so far from the sun that it again shines only by reflected sunlight. It is again a "dirty snowball." Soon it moves so far from the sun that it can no longer be seen. Then after three years, or 60 years, or 105 years, depending on its period (or how far out it goes), we see it again; and the spectacle of its "hour in the sun" is repeated.

Each time that a comet passes close by the sun, it loses a lot of the frozen material that held its rocks together, just as the snowball did on the hearth. No wonder the comet drops material along its path. After several trips around the sun, we would expect so much of the "snow" to be lost that the comet could no longer hang together. Is there any evidence that this is the case?

Some years, a particular meteor shower is especially spectacular. The shower of November 16, 1833, was a fantastic one. Many people thought that the end of the world had come. As many as 200,000 meteors could be counted from one spot in a few hours. Again, on November 16, 1866, and November 16, 1899, there were striking November showers, although the numbers of meteors were less each time. It looks as though somewhere along that orbit there was a dense swarm of fragments (Fig. 17-6) and every 33 years we hit it. In the intervening years we crossed the orbit where the fragments were spread out more thinly. It is possible that this swarm is all that is left of an old comet, with a period of 33 years. Indeed, the swarm was given the name Comet 1866-1.

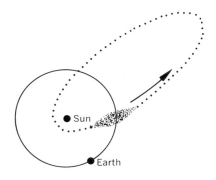

Fig. 17-6

You are probably asking why, if we hit the "carcass" of the comet in 1833, we did not hit the "live" comet in 1800. But at that time the comet was bunched into a "snowball" less than 200 miles in diameter. Our chances of hitting it, of crossing the orbit within this 200-mile stretch, were extremely low. Since loosened debris has spread out much more, perhaps over millions of miles, there is now far more chance of our orbit crossing part of this heavily debris-laden stretch.

As the loosened swarm, traveling in its orbit, nears the sun, it would still, of course, be heated. But in most cases the dispersed, rocky material (individual pebbles) would not glow brightly enough to be visible, just as you cannot see the spread-out fragments as they move close to the sun. Halley's comet, however, continues to be visible; it must have lots of snow and wax left.

Comet material is "boiled off" near the sun and loosened fragments are dropped along the comet's orbit until finally they become so dispersed that they are not bright enough to be seen as they pass close to the sun. Is there any evidence that this is happening to comets we see? We know that the brightness of Comet Encke (period 3.3 yrs) decreased each time it appeared between 1800 and 1950. Its brightness in each case was measured when it was near the sun and "in full flower." This can only mean that the comet was smaller each time it came closer to the sun. It looks as though this comet will be invisible by the year 2000.

Will there be a time when there are no more comets? Or are new comets forming today, keeping up the supply? If so, how are they formed? Where do they come from? Astronomers are attacking these problems today with the patience of Tycho and Kepler.

Additional Reading

PAGE, THORNTON, and L. W. PAGE, eds., *Neighbors of the Earth:* New York, The Macmillan Company, 1965, Chaps. 4-6.

WATSON, F. G., *Between the Planets:* Cambridge, Mass., Harvard University Press, 1956.

Our moon is big, as satellites go. Although there are five larger ones in the solar system (Table 12), they accompany planets much larger than the earth—Jupiter, Saturn, and Uranus. Relative to the size of its planet, our moon is the largest, with a diameter over one-quarter that of the earth. Indeed, the earth-moon system, revolving around its center of mass (p. 71), could be called a "double planet."

Galileo, back in 1609, concluded that the moon is very like the earth. His telescope revealed mountains, craters, valleys, and large, flat areas that he first thought were seas. However, as our view of the moon has gotten clearer and closer (Figs. 18-1 and 18-2), striking differences from the earth's landscape have become evident. What makes the moon so different in appearance from the other half of this "double-planet system"?

It is not the moon's material. The amount of sunlight absorbed by the moon's surface is the same as absorbed by dark rocks or sand on earth; the spectrum of the sunlight that comes to us as moonlight resembles the spectrum of sunlight reflected from rocky material. (And, incidentally, it bears no resemblance to that reflected from green cheese!)

The reason for the moon's different landscape becomes clear if you plot a point on Figure 15-1c, according to the moon's escape velocity (1½ mi per sec) and the temperature of the moon's sunlit side (215°F). You will see that the moon cannot hold an atmosphere, for all the gas molecules would have leaked away. Any molecules now being captured from space by the moon's gravitational attraction, or coming from modern volcanoes or from flows of lava on the moon's surface, would soon be lost.

Planet	Number of known satellites	Diameter range of satellites (mi)	Ratio of diameter of largest satellite to its planet's diameter
Mercury	0	——	——
Venus	0	——	——
Earth	1	2160	0.272
Mars	2	5-10	0.002
Jupiter	12	15-3200	0.035
Saturn	10	100-2600	0.034
Uranus	5	400-1000	0.034
Neptune	2	100?-3000?	0.109?
Pluto	0	——	——

Table 12 Known satellites of the solar system.

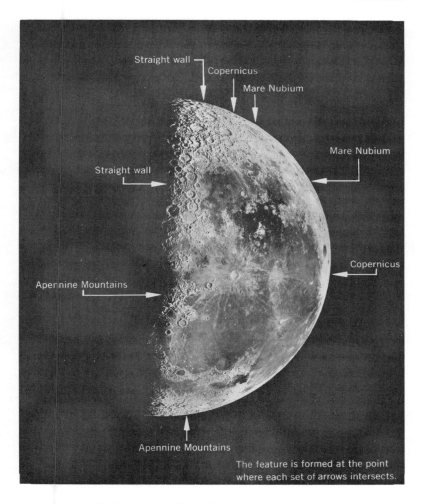

Fig. 18-1 The eastern half of the moon. [Lick Observatory photograph.]

Three observations confirm this prediction and indicate that the moon's surface is practically a vacuum. On the earth, the air scatters sunlight around into the shadow on the night side so that we have twilight for a short time after sunset. Viewed from out in space, as the astronauts have done, this shows up as a gradual shading from light to dark—the twilight zone. In Figure 18-1, however, a sharp line divides the moon's illuminated half from its dark half. Also, when the moon's orbital motion carries it between us and a star, the star's light always blinks out suddenly. If the moon had an atmosphere, the star would dim gradually as the moon's atmosphere came in front of it. In addition, all of the dark lines in the moon's spectrum are those put there by the sun's atmosphere or the earth's atmosphere.

Fig. 18-2 This is part of the first close-up photograph of the moon's crater Copernicus. It was taken on November 22, 1966 by Lunar Orbiter II. Looking due north from the crater's south rim, detail of the central part of Copernicus is shown. Mountains rising from the flat floor of the crater are 1000 feet high, with slopes up to 30°. The 3000-foot mountain on the horizon is in the moon's Carpathian Mountains. Cliffs on the crater's rim are 1000 feet high. From the horizon to the base of the photograph is 150 miles. Lunar Orbiter was 28.4 miles above the moon's surface and about 150 miles due south of Copernicus when the photograph was taken. [National Aeronautics and Space Administration.]

Because there is no air on the moon, there can be no water on the moon's surface. Remember that on earth water boils at a lower temperature on a high mountain than it does at sea level. This is because its boiling point is lower at lower air pressure. If the weight of air above it were zero, as it is on the moon, the water would immediately evaporate into a gas. Then, as a gas, it would disperse into space with the rest of the moon's atmosphere.

On the airless, waterless world of the moon, there can be no wind, rain, or snow. No rivers carry rock material, crumbled by the action of the atmosphere, from high places to low. No ocean waves pound a shore. On the earth, rivers, waves, wind, and glaciers are constantly removing rock material from one place and depositing it in another. They are changing the landscape and have been doing so from time immemorial. This makes the history of the earth difficult to decipher. Part of the record in the rocks is gone, and part of it is buried.

The moon does not have such a "poker face." The record of much that happened to it is probably still written on its surface. Because they are not worked on by the same kind of erosion, features formed in the same way as those on the earth look different. Mountain ranges on the moon, like the Apennine Mountains (Fig. 18-1), have no streams cutting into them, bear none of the marks of glacial erosion, and are not clothed by soil and vegetation. They are generally higher than those on the earth. Prominent cliffs, like the "Straight Wall" (Fig. 18-1), 600 feet high and 80 miles long, are probably the result of slippage from time to time along lines of weakness in the moon's crust. Similar movements on earth cause earthquakes and often form cliffs. Between the periods of movement, however, water and wind efficiently attack these earthly cliffs, and very few are as high or continuous as those on the moon.

The features that really set the moon's landscape apart from the earth's, however, are thousands of craters ranging from a little more than a foot to 150 miles across. The large crater named for Copernicus shows prominently in Figure 18-1; and Figure 18-2 is a detailed view of its interior. The inside walls of large craters like this rise to as much as 10,000 feet above the crater floors, and many have mountain peaks at their centers. Some scientists think that they are volcanoes, and many of the features do resemble Crater Lake in Oregon, an extinct volcano. Most geologists and astronomers agree, however, that the moon's craters were formed by explosions following the impact of fast-moving meteorites of all sizes.

Why should the moon's surface have so many craters, while less than 40 have been found on the surface of the earth? On our planet, erosion immediately begins to remove or bury a crater, and all but the latest ones are destroyed or hidden. On the moon, only the impact of later meteorites can erase an older crater. Then too, we would not expect to find smaller craters on the earth because smaller meteors are burned up as they pass through our atmosphere. In addition, the atmosphere has a slowing effect on incoming meteors. The energy of motion ($\propto mv^2$) of a meteor of mass m, landing on the moon, is much greater than if the same meteor hit the earth, and so it can do more damage as it lands.

Both Mars and the moon, with thin atmospheres and no water, have crater-pocked surfaces (Fig. 14-3) and meteors would be a definite hazard to life there. Meteorites are also the chief means of erosion on Mars and the moon, breaking up the bedrock into fragments of all sizes and hurling them over a wide area. On the earth, however, a meteorite is so rare that it ends up in a museum, and meteorite erosion plays almost no part in sculpturing the earth's surface.

Many craters dot the surfaces of the "maria" (the large, dark plains that Galileo first mistook for seas) that form the features of the "man in the moon." One of these, Mare Nubium, shows clearly in Figure 18-1. These plains may be vast outpourings of lava, but certain observations show that they are not surfaced with solid rock. As the moon enters the earth's shadow during an eclipse, the temperature drops more than 150°C in an hour (as shown by the changing infrared spectrum). This very rapid drop in temperature can be accounted for partly by the absence of the blanketing effect of an atmosphere (p. 109). But the drop is too sharp to be explained fully in this way.

Rocks that have been warmed by the sun take a long time to cool because they are good conductors of heat. They warm to a great depth below their surfaces. But a layer of finely divided material, like dust, conducts heat poorly and warms only at the surface. As a result, it cools quickly. The rapid cooling of the moon's surface suggested that it is covered with a layer of fine dust. Photographs taken in 1957 by NASA's Lunar Lander confirmed this. It seemed likely that the dust is volcanic ash or tiny fragments of meteorites and bedrock broken up by the impacts of meteorites, or that it came from both these sources.

On July 21, 1969, men explored the lunar landscape that Galileo's telescope had first revealed. Col. Edwin Aldrin, Jr., and Neil Armstrong landed the lunar module of the Apollo-11 spacecraft in the "Sea of Tranquillity" and left their footprints in the lunar dust. They collected and brought back samples of it. Some of the dust particles proved to be angular rock fragments, but, surprisingly, over half the particles were glass "beads," spheres of glass less than 2 mm. in diameter. They are thought to be spatterings of meteorite-pelted rock that hardened as they were whirled high above the lunar surface.

The Apollo-11 astronauts also collected rock specimens, bringing back coarse- and fine-grained rocks which resemble terrestrial ones that have hardened from molten magma or lava. In addition, the Apollo-12 astronauts, who landed in the "Sea of Storms" (Figs. 18-1, 2) on November 19, 1969, found rocks made up of angular, pebble-sized fragments of lava-like rock, bound together by glass.

It will take years to learn all that the rocks and dust can tell us about the moon. How much of the "lava" and glass came from volcanic outpourings is not certain. Surely some are rocks melted by the terrific impact of meteors, and hardened again into rock or glass. Radioactive studies have shown that one specimen of lava-type rock brought back by the Apollo-12 astronauts, Alan Bean and Charles Conrad, Jr., is 4.6 billion years old. This great age suggests that it may be an unaltered piece of the moon's original crust.

The astronauts who landed on the moon's surface found it bleak and lifeless, pockmarked by craters of all sizes, strewn with boulders large and small, and covered with dust. They found no evidence of living things, and did not expect to. They were able to exist on the moon only because of their life-support systems brought from earth, attached as back packs to their space suits.

During the moon's two-week-long day, the bright side heats to above 212° F, the boiling point of water on earth; on the night side the temperature is down to −280° F. There is no air to breathe or water to drink. Our earth's atmosphere absorbs most of the ultraviolet light from the sun, but it arrives on the airless moon in full strength. Eventually it would kill all living things not shielded from it.

The earth may have captured chunks of rock too small to be seen as they orbit the earth, but until October 4, 1957, the moon was earth's only known satellite. Then the first man-made moon, Sputnik I, was launched, and now there are over 1000 of them in orbit. Astronomers eagerly await the information brought back by these satellites and by space probes, which they cannot get with earth-based telescopes because of distance and because only visible light and long radio waves can come in through the atmosphere. Other wavelengths could tell us much about the temperatures and composition of the stars. Also, air currents in the earth's atmosphere make stars "twinkle" and smear photographs made with telescopes, causing loss of fine details. The Orbiting Astronomical Observatory, launched in 1967, is as great an improvement over ground-based telescopes as Galileo's telescope was over naked-eye observations.

Satellites, probes, and space travel would not be possible if astronomers had not worked out the geography and movements of the solar system—the distances of the planets, their locations at any given second, their sizes, and the conditions on each one. They have provided the Space Age with maps, timetables, and tourist information. Newton's laws assured men that a spaceship could be put into orbit; and when man's wealth, leisure, and engineering skill became equal to the task, it was done.

In Chapter 3 (pp. 59-60), we followed Newton's line of reasoning that if a stone could be given a great enough forward speed, it would completely encircle the earth, always falling toward the earth but never reaching it. It would be in orbit like the moon, on a curved path around the earth's curved surface. The moon had somehow gotten a sufficient sideways speed (and so had the planets, as shown in Fig. 8-2). The problem was to give enough sideways speed to a satellite.

In the 1930's, rockets began to be developed. Guided rockets were first used by the Germans in World War II. These rockets were able to push an explosive shell in the right direction and to keep on pushing it for hundreds of miles until the fuel gave out. The acceleration continued so long that tremendous speeds were reached. The early rockets, or missiles, could travel for hundreds of miles before gravity brought them down. Later missiles were accelerated to speeds high enough that when a man-made moon was substituted for the shell, it could be put into orbit.

How does the missile keep on pushing? Inside a missile, hot gases under tremendous pressure are formed by the burning of high-explosive fuel and then forced out through a tailpipe. Released from the high pressure, they stream out behind the rocket. This pushes the rocket forward, which would not surprise Newton a bit. He reasoned (p. 58) that as the force of gravity pulled the apple down, it also pushed the earth upward. Since $F = ma$, the tiny apple was accelerated downward much more noticeably than the massive earth was affected. Newton felt that there was no reason why the force, caused by the masses of both earth and apple, would only affect one of the masses, the apple. Besides, he had seen similar effects from other sorts of forces. You have, too. If you dive from a floating rowboat, the boat moves in the direction opposite your dive. When you fire a gun, the bullet moves forward and the gun kicks back against your shoulder with equal and opposite force. As the stream of gas from a rocket or jet plane shoots backward, the rocket or plane moves forward. Newton therefore concluded, "To every action, there is an equal and opposite reaction." Of course, he was extrapolating beyond experience, but satellites orbiting the earth show that he had the right idea.

In launching a satellite, several rocket motors are used. These are called *stages*. In order to reduce the weight and the amount of fuel needed, each stage is dropped as its fuel is consumed, and the next one is fired. A large rocket usually gives the first push. When the rocket system reaches the top of its flight, the remaining stages are fired in turn, all aimed roughly parallel to the earth's surface.

The farther out a satellite is, the less is the pull of gravity on it. Therefore, satellites with larger orbits need less forward speed to keep them from falling back to earth. The early satellites orbited at distances of several hundred miles

Fig. 18-3 Paths of the empty balloon satellites Echo I and Echo II, the brightest man-made moons in the sky, are recorded by a time-exposure photograph in front of the Minuteman statue in Concord, Massachusetts. Echo I, launched in 1960, is orbiting from lower left to upper right. Echo II, shot into orbit in 1964, was tracked from right to left. An airplane left its trail at the lower right. Star tracks as the earth rotated are also caught in this 12-minute time exposure made by Charles Hanson early in 1966.

above the earth, where forward speeds of about 18,000 miles per hour were necessary to keep them in orbit. Their periods were about 90 minutes. Later satellites, forced higher, orbited more slowly. Most of the early satellites, some of the later ones, but none of the manned ones, were close enough or large enough to be seen (Fig. 18-3) as they moved in their orbits against the background of the stars. If you live in Boston, New York City, Philadelphia, or Washington, D. C., you can phone Dial-A-Satellite to find out if any are visible tonight from your area.

In the zone where most of the satellites travel, the air molecules are about 12 miles apart (as compared to one-billionth of an inch apart at sea level). Nevertheless, the resistance of even this extremely thin air eventually slows a satellite down, and the pull of gravity is able to draw it nearer to the earth. At last it slips down to a level where the atmosphere is much more dense, and then friction with the air tears it apart and sets it ablaze. However, Explorer I, the second U.S. satellite, which reaches a distance of 4000 miles from the earth on each orbit, is expected to have a lifetime of 200 years.

Space probes, which do not orbit the earth, must be designed and fueled to go faster than 25,000 miles an hour, the earth's escape velocity. Like other members of the solar system, however, they do travel in orbit, but their orbits are around the sun. In a sense, they are new minor planets. Two of these are the Mariner Venus probe launched in 1962, and the Mariner IV Mars probe that took the photograph in Figure 14-3. Lunar Orbiter II was guided into an orbit around the moon. It is a satellite's satellite, something not yet found occurring naturally in the solar system. Ranger IX, that sent back Figure 18-1, crashed into the moon rather than orbiting around it.

Contrary to what many people think, astronauts in orbit are not weightless because they are so far from the earth. At 400 miles, or 4000 miles, or 40,000 miles, their weight would be less, of course, than on earth (p. 69) but it would not be zero. The weight that you feel on your feet when you are standing up, or on your arms when you hang from a horizontal bar, or on your back when you lie down, is the force opposing the gravitational attraction of the earth on your body. You exert a force on the floor as gravity accelerates you toward the center of the earth, and the floor exerts an equal and opposite force on you. (If it did not, you would go crashing through.) This force opposing your weight is measured in pounds on a bathroom scale. If you took such a scale along with you in an artificial satellite orbiting the earth, it would read zero when you climbed on to it to get weighed. (No one has yet done this, partly because the astronauts can feel their lack of weight quite easily, and because the reason for it is so simple.) But, as we said, the astronauts are not weightless because they are so far from earth. They are weightless because they are in orbit.

Some of the effects that result from weightlessness are confusing to the astronaut: If he tries to pour water from a cup, it will not pour because the cup is falling just as fast as the water. If he tries to light a candle, there is no flame because when both hot air and cold air are falling together, hot air does not rise in cold air to make the candle flame.

Of course, there is a small gravitational attraction between the astronaut m and the center of mass of the satellite M, since $F = GmM/r^2$, but the satellite's mass is so small that this force can hardly be detected. On the moon, a much larger satellite, M is about $1/80$ the mass of the earth. So if you stood on the moon's surface, you could feel its attraction as about $1/5$ the weight you have on the earth's surface, for the moon's radius is about $1/4$ that of the earth's.

If the spacecraft's rocket motor is on, the astronaut feels the force of its push. This force can be up to eight or ten times his normal earth weight during blast-off or in reentry. In the same way, an astronaut on a space probe bound for the moon, Mars, or Venus, will feel some part of his weight when the space probe is turned or slowed down by rocket motors; but during ordinary orbiting with rockets off, he is weightless.

Additional Reading

ARMSTRONG, NEIL, COLLINS, MICHAEL, ALDRIN, EDWIN E., JR., FARMER, GENE, and HAMBLIN, D. J., *First on the Moon:* Boston, Little, Brown and Company, 1970.

MOORE, PATRICK, *Guide to the Moon:* London, W. W. Norton Ltd., 1953.

OVENDEN, M. W., *Artificial Satellites:* Baltimore, Penguin Books, 1960.

Photographs of the sun made through a telescope look like Figure 7-6. They show a smooth, round disk called the *photosphere* (meaning sphere of light), marred only by sunspots (p. 53). The photosphere looks hard and solid. However, from the Planck curve (p. 100), Stefan's law (p. 96), and Wien's law (p. 99), it is known that the temperature of the sun's surface is about 6000°C. With such a high temperature the sun's material must be entirely gaseous rather than solid or liquid. This is shown by laboratory experiments with material at high temperatures and is confirmed by the changing shapes of the sunspots on a series of photographs.

Although spectra show that the gases in the sunspots are about 1500°C cooler than those surrounding them, they are still very hot. Doppler shifts show that the gas in a sunspot is moving about 2 miles per second, flowing outward and upward from the center of the spot and downward and inward toward its center. Sunspots range from specks 500 miles across to giant whirlpools more than 500,000 miles across — up to $1/15$ the sun's diameter. Sunspots often come in groups, and the total number seen each year varies in a cycle of about 11 years. Over a hundred sunspots, for instance, were seen in 1948 and in 1959, while very few were observed in 1954 and 1965. No one has explained this 11-year cycle, although it has long been known that when there are many sunspots, radio and telegraph communication on earth is poor. Also, the earth's weather shows changes during each sunspot cycle which are repeated during the next cycle.

Aside from the sunspots, the photosphere appears smooth and featureless in Figure 7-6. However, we can see from photographs taken with a telescope in a balloon 80,000 feet above the earth, above the blurring effect of earth's atmosphere, that the photosphere has a mottled surface resembling rice grains. The smallest "grains" are about 300 miles across. Doppler shifts in the spectra of the individual grains show that they are columns of hotter gases, rising from layers below the sun's surface. As the rising gas reaches the top of the photosphere it spreads out and sinks down again. The darker boundaries of the grains are the cooled gases sinking back into the photosphere. (These are only about 75°C cooler than the centers of the grains.) The vertical motions have speeds of 1 to 2 miles per second, and individual grains persist for only a few minutes. The whole surface of the sun thus appears to be in activity, gases violently bubbling up and sinking down continually.

When a spectrograph is attached to the telescope, the spectrum of the light coming from the photosphere can be photographed. This is a continuous spectrum, crossed by many dark, narrow absorption lines. This means that the photosphere is emitting light of all wavelengths, but when this light passes through the sun's atmosphere, some of the colors are absorbed, leaving dark lines across the spectrum. In Figure 7-6 we are looking at the sun through its atmosphere.

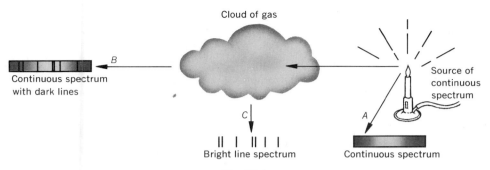

Fig. 19-1

The absorption lines are labeled for various chemical *elements*. They are the lines that would appear in the continuous spectrum of light shining through a gas consisting of *atoms* of these elements. The dark lines added to the spectrum of sunlight by the cool atmosphere of Venus are broader than these lines and occur at different wavelengths. They are the lines put in by *molecules* of carbon dioxide (CO_2). A compound, like CO_2, is made up of elements; each molecule of carbon dioxide contains one atom of carbon and two atoms of oxygen. The conspicuous dark lines put in the spectrum of sunlight by the earth's atmosphere are those of oxygen molecules (O_2), each made up of two atoms of oxygen.

The lines in the sun's spectrum show that atoms, rather than molecules, make up the sun's atmosphere. This is to be expected, for laboratory experiments have shown that at such a high temperature molecules break up, releasing their atoms.

In Figure 19-1 if we look along line *A*, directly at the Bunsen burner, we see a continuous spectrum. The spectrum of the sun would likewise be simply a continuous spectrum if we could strip away the sun's atmosphere. When we look at the burner through the cloud of gas (along line *B*, Fig. 19-1), we see the continous spectrum crossed by dark absorption lines. This corresponds to the view of the sun that we have in Figure 7-6. If we look along line *C* in Figure 19-1, we do not see a continuous spectrum; we see only bright lines. These bright lines are at the same wavelengths as the dark lines we saw when we looked along line *B*. For instance, if the gas were made up of hydrogen atoms, we would see dark lines (spectrum *B*) and bright lines (spectrum *C*).

It looks as though hydrogen atoms selected just these wavelengths of light from the variety being sent out by the burner (or the sun) and absorbed them. The hydrogen atoms must also be giving out light of just these same wavelengths, because if we look along line *C*, we see bright lines in these same positions in the spectrum.

This is puzzling. If the hydrogen atoms are giving out light of the same wavelengths that they absorb, why do we see any dark lines at all? Why doesn't the reemitted light fill in the gaps and make us none the wiser that certain wavelenghts were stolen and then returned? The reason is that the light from the flame is coming to our eyes along one direction, *B*. On the way, the hydrogen atoms take out

141

most of the light of the four definite wavelengths, and they keep on doing it as long as the light is shining. At the same time, they are reemitting light of these same wavelengths in *all* directions. The result is that only a small fraction of the reemitted light happens to travel in the same path as the light from the flame (Fig. 19-1). Hence, while the dark lines are not quite as dark as they would be if the atoms didn't reemit at all, they are still dark in comparison to the rest of the spectrum.

In order to see the sun's bright-line spectrum, we must look along a direction like *C* in Figure 19-1, where we don't see the continuous spectrum at all. Can we look at the sun along a direction corresponding to line *C*? We would expect to find the sun's atmosphere extending beyond the sun itself, and if we could look at it there, we should see bright lines. The photosphere ends sharply; there appears to be nothing beyond it. However, in the seventeenth century during an eclipse of the sun (Fig. 3-3a), several observers described, for the first time, a narrow red streak or "fringe" around the edge of the sun for just an instant before and after the shining disk of the sun was completely covered (Fig. 19-2). This is the sun's atmosphere just above the photosphere. It is called the *chromosphere* (sphere of color). It could only be seen when the bright photosphere was cover-

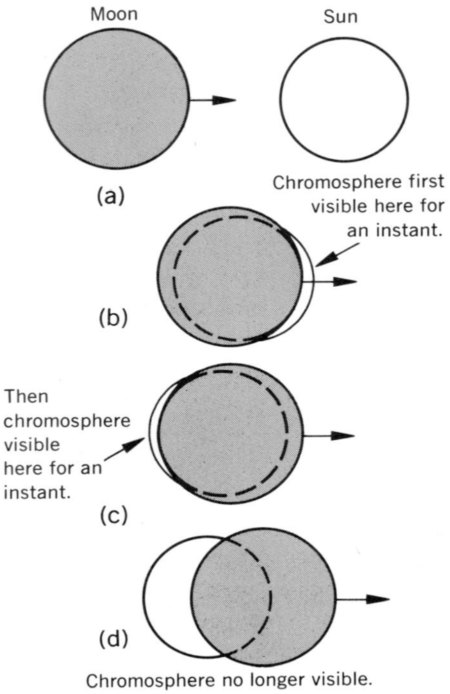

Fig. 19-2 The moon's average angular diameter is slightly larger than the sun's. (a) The moon, moving in its orbit around the earth, nears the sun in the sky. (b) The moon passes between the earth and the sun, eclipsing the sun. As the forward edge of the moon lies, for an instant, along the edge of the photosphere, the chromosphere is visible. (c) As the moon moves on, its other edge, for an instant, lies along the other edge of the photosphere, and the chromosphere is briefly visible. (d) The moon has moved on, exposing part of the photosphere, and the chromosphere is no longer visible.

ed by the moon; otherwise the intense light of the photosphere overwhelmed its fainter light. (Today, special instruments called spectroheliographs enable astronomers to photograph the chromosphere at any time.)

In 1868 the spectrum of the chromosphere was first photographed and, as you would expect, was found to be made up of bright lines. The reddish color of the chromosphere is due to one of its strongest emission lines, that of hydrogen at 6563 A.

Violent storms in the chromosphere send up tongues of glowing gas, called *prominences* (Fig. 19-3). Curtainlike, they arch over great distances, as much as 100,000 miles above the surface of the sun, their gases moving at speeds of

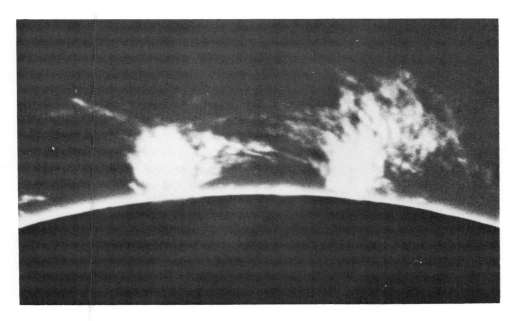

Fig. 19-3 A view of a solar prominence taken by Donald H. Menzel. [Courtesy Sacramento Peak Observatory.]

100 miles per second or more. In other prominences, material can be observed rapidly moving downward, back into the sun. The cause of prominences is unknown.

The chromosphere extends only about 5000 miles above the photosphere (as indicated by its angular width). It merges into the outermost part of the sun's atmosphere, the *corona* (Fig. 19-4). Like the chromosphere, the corona was first observed only during total eclipses, but unlike the chromosphere, it had been seen during eclipses for centuries. Because it is a wider band, extending for more than a million miles beyond the photosphere, it is visible for a longer time as the moon moves across the sun. Like the chromosphere, its usual invisibility is due to the overpowering light of the photosphere. The outer part of the corona shines by sunlight reflected from tiny particles of dust. Its inner part, however, shows about two dozen bright

Fig. 19-4 The solar corona, photographed by Georges Van Biesbroeck in Brazil at the eclipse of May 20, 1947. [Yerkes Observatory photograph.]

143

emission lines. These lines indicate temperatures of almost a million degrees centigrade! It seems odd, indeed, that the outermost atmosphere of the sun, so near the extreme cold of outer space, should be its hottest part. Here is something for astronomers and physicists of the future to explain!

Atoms in a hot gas radiate bright, isolated lines in the spectrum, a different set for each atom. The glowing filament of an electric light bulb made of tungsten, however, radiates a continuous spectrum. So does a red-hot poker, so does a piece of white-hot iron, so does any glowing solid. Every visible wavelength is bright; photons of all different amounts of energy (all colors of light) are being emitted. How can this be?

In a gas, like the sun's corona and chromosphere, the atoms are far apart. Neighboring atoms have little influence on each other. Most of the time an atom moves around just as if there were no other atoms there. The atoms in a solid body, like the tungsten in the light-bulb filament, are never free of closely packed neighboring atoms. An outermost electron is almost as close to the nuclei of several neighboring atoms as it is to the nucleus of its own atom, and it is pulled by the positive electric charges of these other atoms as well as by the charge in its own nucleus. As a result, there are many orbits of many atoms into which it can jump, and these orbits are changing all the time. For any electron there are many different photons of energy which it can absorb, many different photons that permit a jump somewhere. An electron is not limited to radiating in a definite set of wavelengths. It can absorb a whole variety of photons, pass part of its energy to other electrons, and emit completely different photons. At any moment, the billions of electrons in a solid are emitting billions of different wavelengths. The result is that a solid body radiates a continuous spectrum, and every possible wavelength is represented.

But at 6000°K the sun's photosphere cannot be a solid. The way that the sun rotates also shows that it is not solid. Doppler shifts in spectra taken near the edge of the photosphere (at positions corresponding to A and B in Fig. 16-2) give the rotational speed, in miles per hour, of a point on the sun's equator. Since the sun's circumference is known from its distance and angular diameter, this tells us that the period of rotation of the equatorial regions is 24 days, 16 hours. However, similar observations show that at points north and south of the sun's equator the rotational period is longer. At solar latitude 75°, for instance, it is 33 days. A solid body, like a planet, cannot rotate like this. A point on the earth rotates at a slower hourly speed the farther it is from the earth's equator, and every point on the earth completes one rotation in 24 hours.

How can the gaseous photosphere (Fig. 19-5) radiate a continuous spectrum? At any point deep below the sun's surface, a greater weight of gas lies above it. This weight compresses the gas below into a smaller volume; it increases the density of the gas, the mass per cubic centimeter. In other words, there are more atoms in each cubic centimeter which means that atoms are closer to one another than in the gas above the surface of the photosphere. Like any gaseous atoms, they are in motion; and at higher temperatures they move faster and bump into each other sooner. Deeper down, they have even less room to move about. At the density of the photosphere, the effect is the same as if each atom were locked in place very close to other atoms, as in a solid. As a result, the photosphere (and other hot, compressed gases) behaves very much as a glowing solid: It absorbs and emits light of all wavelengths and produces a continuous spectrum.

The mass of the sun, its diameter, its average density (mass/volume), its surface temperature, and the amount of radiation that it emits per second are known. So is the structure of atoms and the behavior of material at the highest pressures and temperatures obtainable in the laboratory. It is obvious that the sun's gravity

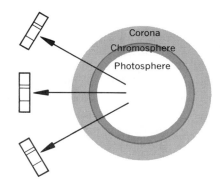

Fig. 19-5 Cross-section of the sun. The photosphere radiates a continuous spectrum. The chromosphere and corona absorb certain wavelengths, producing absorption lines, as shown schematically on the left.

is pulling material toward the center, tending to collapse the whole mass into the smallest possible volume. This inward pull is balanced by two forces: the outward pressure of radiation and the outward movement of gas molecules toward less dense regions. Both of these increase with increasing temperature; they increase toward the sun's center—but so does the opposing pull of gravity. At any point in the sun, these opposing forces must balance; otherwise the sun would change in size. Taking into account all these facts, mathematicians have calculated how the pressure, temperature, and density of the sun increase with depth. They tell us that the density near the sun's center must be a hundred times that of water, and the temperature there must be about 20 million degrees.

Each second, 6×10^{27} calories of energy are pouring out from the photosphere. If the sun's interior were not producing energy at this same rate, it would cool off. Geological evidence indicates that the sun has been delivering radiant energy to the earth at about this rate for the past 4½ billion years.

How does the sun produce this vast amount of energy and keep on doing it for billions of years? The earliest answer to this question was simple (and wrong); until about 1800 men thought that the sun's material is burning, like wood burns in a fireplace. They thought that the rapid union of oxygen (O_2) and another substance, like the carbon (C) in the wood, gives off heat and light in the sun and produces carbon dixoide (CO_2); and that combinations of other elements, like chlorine and hydrogen, also produce its heat and light. But eventually all the wood would be gone (locked up in CO_2 molecules), the fire would go out, and the fireplace would be empty except for a pile of ashes that will not burn. The sun, shining 4.5×10^9 years ago, would have consumed all its materials long before now if they were simply on fire. Looked at another way, the material of the sun would be all used up 2000 years from now. Besides, the sun is too hot to burn. The carbon could never unite with the oxygen, or the chlorine with the hydrogen. The atoms are moving at such high speeds that such chemical combinations are impossible.

The Sun

In the 1850's two scientists, one English and one German, presented another idea. The sun shines, they said, because it is shrinking. As it shrinks, its particles fall inward, and the energy of this falling motion is converted into heat energy. The gravitational force of the sun is so strong that a shrinking of 150 feet per year could provide all the energy radiated. This is a reasonable theory until you look at it more carefully. In the 350 years that man has been observing the sun with telescopes, this would amount to a 20 mile change in diameter—much too small to be measured. Tracing it backward, however, we find that 20 million years ago the sun would have been larger than the earth's orbit. (At that time it would have had to shrink 35,000 miles each year to keep shining.) In rocks laid down at that time on earth are the bones of elephants, horses, and apes; they could scarcely have survived inside the sun, nor could their bones have been covered by sand and silt carried by liquid water, like the sediments laid down in streams and lakes today.

Up until 1908, scientists had thought of mass and energy as two different things. Then Albert Einstein showed that mass is nothing more than a special kind of energy. He showed that the relation between mass and energy can be expressed by the equation $E = mc^2$ (where the energy E is given in ergs, m is the mass in grams, and c is the speed of light, 3×10^{10} cm per sec). If a one-gram mass could be converted into energy, it would produce 2×10^{13} calories: $E = (1)(3 \times 10^{10})^2 = 9 \times 10^{20}$ ergs $= 2 \times 10^{13}$ calories. This is the energy that we would get if a one-gram piece of wood (or any other substance) could be entirely converted to energy. It would be sufficient to keep a 100-watt bulb lighted for 30,000 years. If, on the other hand, we merely burn a one-gram piece of carbon, using 2.6 grams of oxygen to do it, we get 3.6 grams of CO_2 and about 10^3 calories of energy. No mass is lost in the burning; the atoms of carbon and oxygen are still intact. But they are combined in the molecules of CO_2, off to become a gas in the earth's atmosphere.

As early as 1917 it was suggested that nuclear changes, and the conversion of mass into energy that accompanies them, might account for the sun's long-continued and huge energy output. The astronomers asked themselves whether hydrogen nuclei (protons), which make up 80% of the sun's mass, could join together to produce heavier nuclei, turning some matter into energy as they do so. Since hydrogen and helium are so common in the sun, Hans Bethe and Edwin Salpeter of Cornell University, in the late 1930's, worked out nuclear reactions that convert hydrogen into helium.

At the high temperatures (6000°K) near the surface of the photosphere, the tremendous speeds of the colliding hydrogen molecules cause them to break into hydrogen atoms (H). At much higher temperatures, as in the chromosphere and inner corona, the atoms are moving so fast that the hydrogen atom's one electron is knocked right out, and the atom becomes an ion (H^+). Down in the interior of the sun, where astronomers estimate that the temperature is as high as 20 million degrees, the speeds are so great that the charges cannot repel each other any longer. So nuclear reactions do take place. Hydrogen is converted into a smaller mass of helium; the lost mass is converted into energy.

The result is a loss of mass of 0.02803 atomic-mass units; for the weight of the helium nucleus is 4.00389, which is less than the weight of the four hydrogen units ($4 \times 1.00813 = 4.03252$) that formed it. Thus, of every 4.03252 grams of hydrogen that is converted to 4.00389 grams of helium, 0.02863 gram (about 0.7%) is converted into energy. According to Einstein's equation, $E = mc^2$. Thus, we can calculate $E = (0.0071)(3 \times 10^{10})^2 = 6.4 \times 10^{18}$ ergs. Each gram of hydrogen in the sun can produce 6.4×10^{18} ergs of energy.

Direct measurements have shown that the sun emits 3×10^{33} ergs (or 6×10^{27} calories) per second. Since each gram of hydrogen produces 6.4×10^{18} ergs, this means that 5×10^{14} grams of hydrogen ($3 \times 10^{33}/6.4 \times 10^{18}$) are being used up each second. The sun is losing this much hydrogen each second. Since there are 3×10^7 seconds in a year, it is losing ($5 \times 10^{14})(3 \times 10^7$), or 1.5×10^{22}, grams of hydrogen each year. How long can this keep up? The mass of the sun is 2×10^{33} grams; and of this, 80% is hydrogen. However, only hydrogen fairly deep in the sun is hot enough and dense enough to permit the proton-proton reaction to go on. It seems reasonable to consider that 50% of the sun's mass (10^{33} grams) is the amount of hydrogen which can be used in the proton-proton reaction. Since 1.5×10^{22} grams are used each year, it looks as though the sun could continue to shine for 6.6×10^{10}, or 66 billion, years (because 10^{33} grams divided by 1.5×10^{22} grams per year $= 6.6 \times 10^{10}$ years). Long before the hydrogen is gone, the sun would change, partly because the chance of protons colliding with other protons would drop. Nothing much should change before only a quarter of the hydrogen is used up—in 16 billion years.

Thus, nuclear reactions within the sun can provide us with over 10 billion years' more of sunshine like we have now. There is no problem in explaining the features in rocks 450 million years old, that indicate that the earth was then much like it is today.

It seems likely that the other stars also produce their light by nuclear reactions. We would expect these other, more distant suns to be made up of hot gases, too. If the temperature, composition, and size of a star are the same as those of the sun, we would expect that the proton-proton reaction would be operating. If the conditions of other stars are quite different, then other nuclear reactions may power their furnaces. To find out, we must learn more about the other stars.

Additional Reading

GAMOW, GEORGE, *A Star Called the Sun*: New York, Viking, 1964.

MENZEL, D. H., *Our Sun*: Cambridge, Mass., Harvard University Press, 1960.

PAGE, THORNTON, and L. W. PAGE, eds., *The Origin of the Solar System:* New York, The Macmillan Company, 1966, Chaps. 2 and 3.

PAYNE-GAPOSCHKIN, CECILIA, "The Sun" in *Astronomy* (Samuel Rapport and Helen Wright, eds.): New York, New York University Press, 1964.

Long before astronomers had telescopes or had worked out the temperature and density in the sun, or what makes it shine, they guessed that the sun is a star. On a clear dark night you can see about two thousand stars, and none of them look anything like the sun. Nevertheless, Copernicus and Tycho Brahe reasoned that the sun would look like a star if it were about one hundred thousand times (10^5) farther away, at a distance of 10^5 AU instead of 1 AU. Then it would seem $(10^5)^2$, or 10^{10}, times fainter than it does now — or about as bright as the stars of the Big Dipper.

When Henderson, Bessel, and Struve made the first measurements of stellar parallax (p. 64) back in 1838, they found that during a year the bright star Alpha Centauri shifted only 1.5″ (one and a half seconds of arc), 61 Cygni, visible only through a telescope, shifted through 0.6″, and the very bright star Vega shifted through 0.25″. This showed that the early guess on star distances was about right for these three stars. As shown in Figure 20-1, if the yearly shift is 2″, the angle p (the parallax) is 1″. It can be calculated by geometry that if angle p is 1″, the distance from the earth to the star is 206,265 AU (a star with a parallax of 1″ is 206,265 times farther from us than the sun).

The distance 206,265 AU is called a *parsec* because a star at that distance has a *par*allax of one *sec*ond. Since parallax is inversely proportional to distance (Fig. 5-3c, d), the distance R of a star (in parsecs) is given by the equation $R = 1/p$ with p measured in seconds of arc. The three stars measured in 1838 are between about 2.7×10^5 and 1.7×10^6 times farther from us than the sun, as shown in Table 13.

The stars are so far away that distance alone would make them look like points of light, rather than brilliant disks like the sun. Just about the smallest parallax that can be measured with certainty is 0.05″, which corresponds to a distance of 20 parsecs (4×10^6 times our distance from the sun). Only about a thousand stars

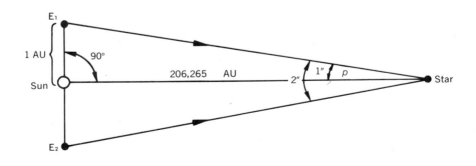

Fig. 20-1

	Alpha Centauri	61 Cygni	Vega
Observed yearly shift (seconds of arc)	1.5	0.6	0.25
Parallax, p (seconds of arc)	0.75	0.3	0.13
Distance, R, in parsecs ($R = 1/p$)	$1/0.75 = 1.3$	$1/0.3 = 3.3$	$1/0.13 = 8$
Approximate distance in AU ($206{,}265 \times$ distance in parsecs)	268,150	680,675	1,650,120

Table 13 Parallax and distance of Alpha Centauri, 61 Cygni, and Vega.

have parallaxes large enough to measure, so we would guess that most of the millions of faint stars visible through the largest telescopes are even farther away.

Light, traveling at 186,000 miles per second, takes eight minutes to come to us from the sun. The light that reaches your eyes tonight from Sirius (Fig. 2-6c) left that star eight years ago. (Sirius, at a distance of 2.6 parsecs, is 4.7×10^{13} miles away, and light travels 5.88×10^{12} miles in a year.) You are not seeing Sirius as it looks tonight, but as it looked eight years ago. The distance of Sirius is often given as eight *light-years* (rather than as 2.6 parsecs). Like the parsec, the light-year is a unit of *distance* (one parsec = 3.26 light-years), although it does sound more like a unit of time.

If the stars were all just like the sun, we would expect that the nearer a star is (Table 13), the brighter it would look. Vega, at four parsecs, is so bright that she is called "Queen of the Summer Sky." Sirius, at 2.6 parsecs, is the brightest star visible from the United States. Yet 61 Cygni, 0.9 parsecs nearer than Sirius and 2.3 parsecs nearer than Vega, is so dim that it can be seen only through a telescope! This suggests that the stars cannot be all alike. Some must be emitting more radiant energy than others; some must be more *luminous*. If we could place Sirius and Vega at the same distance from us, Sirius would not appear as bright as Vega. Would the sun and Alpha Centauri? Do the stars differ greatly in their light-giving power?

In order to be sure, astronomers first had to know the relative brightness in the sky of stars whose distances they knew. By the time the first stellar parallax was measured, the relative brightness of many stars had been determined. If you go out and look at the stars, you will see that they differ greatly in brightness. You can estimate that star *A* is about twice as bright as star *B*, and star *C* about half as bright as star *B*, and so on. Back in the second century B.C., the Greek scientist Hipparchus did just this. He listed about a thousand stars according to their brightness, and divided them into six "magnitudes." In the first magnitude he placed the brightest stars; the faintest that he could see were in the sixth magnitude, and the other stars were assigned to magnitudes between, according to their brightness.

149

The Stars as Other Suns

Sir William Herschel, the discoverer of Uranus, thought of a method to make these measurements of relative brightness more exact. The light-gathering power of a telescope is proportional to the working area of the lens or mirror (Figs. 7-1 and 11-1) which, in turn, is proportional to the square of its radius. With a lens diameter of 8 inches, Herschel might see two stars, the dimmer one barely visible. When he covered more of the lens, reducing its diameter to 4 inches, the dimmer star could not be seen at all, and the brighter star became barely visible. Since only one-fourth as much light was entering the telescope, he reasoned that four times as much light is arriving from the brighter star than from the dimmer one. He and his sister Caroline made thousands of measurements of relative brightness, as did many astronomers who came after them. Later, photoelectric cells were used for even greater accuracy.

Table 14 shows the distances R of eleven of the brightest stars and their relative brightnesses b. If these stars were all sending out the same amount of light, their observed brightnesses would be proportional to $1/R^2$. Clearly, they do not. The stars differ widely in *luminosity,* as we suspected when we compared the distances and brightnesses of 61 Cygni and Sirius.

If all the stars had the same luminosity, a star's brightness would be proportional to $1/R^2$. If they were all at the same distance, a star's brightness would be proportional to its luminosity ($b \propto L$). Thus, $b \propto L/R^2$, and by algebra, $L \propto bR^2$. As you know, an equals sign can be substituted for $\propto$, if a constant k is added. If we describe a star's brightness in terms of the sun's brightness, and use R in AU, then $k = 1$, and we can solve for L in terms of the sun's luminosity: $L = bR^2$. For instance, a star that appears 10 million millionths (10^{-11}) as bright as the sun in the sky, and is a million times as far away (10^6 AU or about five parsecs), has a

Star*	Brightness (Pollux = 1)	Distance in parsecs (R)	R²	1/R²
Sirius	10.8	2.7	7.3	0.137
Alpha Centauri†	2.94	1.3	1.7	0.588
Arcturus	3.08	11	121	0.0083
Vega	2.80	8	64	0.0156
Capella	2.78	14	196	0.0051
Procyon	2.05	3.5	12.3	0.0813
Achernar†	1.82	20	400	0.0025
Altair	1.43	5.1	26	0.0385
Aldebaran	1.32	16	256	0.0039
Pollux	1.00	12	144	0.0069
Fomalhaut	0.97	7	49	0.0204

Table 14 Distances and observed brightness of some conspicuous stars.
* For location in the sky, see Figures 2-6a through d.
† This star is not visible from most of the United States.

luminosity L of $(10^{-11})(10^6)^2$, or 10, suns. Its luminosity is 10 times that of the sun. It is easier to use parsecs for the large distances R to stars, and to use the fact that if the sun were 10 parsecs away, it would appear 100 times fainter than the brightest stars in the sky (0-magnitude stars) of brightness b_0. Using b_0 as the unit of brightness, the luminosity in "suns" is simply $L = b_0 R^2$. Since the sun's luminosity is 6×10^{29} calories per second (p. 144), we need only to multiply L by this number to get the luminosity in calories per second from b and R. Astronomers prefer to use the larger unit, "suns."

Luminosities vary a great deal. This could mean that the stars vary in size, since at equal temperatures a large object radiates more heat and light. Or it could mean that they are all the same size but have different temperatures, since the higher the temperature, the more light and heat is radiated by objects of the same size. Or it could be both. Since the stars appear only as points of light in even the largest telescopes, we cannot measure a star's size by its angular diameter, as we did for the planets. However, on a clear night you can easily see that the stars are not all of the same color. Sunlight, seen reflected off the moon, is much whiter than the light of Antares (Fig. 2-6b), which shows a reddish tinge. Sirius and Vega look much bluer. The sun's photosphere looks yellowish; its Planck curve shows that it is giving out most of its light in yellow-green wavelengths. This suggests that a study of the spectra of the stars will tell us about their temperatures. The Planck curves will be different for stars of different temperature.

Astronomers have been studying stellar spectra for 150 years, ever since Fraunhofer (p. 116). They have found, for instance, that a bright yellowish star like Capella (Fig. 2-6c) gives a spectrum very like the sun's spectrum, with most of the same absorption lines. Other stars have quite different spectra. Some, like Sirius, show only the lines of hydrogen, some show only the lines of helium, and some show the broad bands of molecules. Does this mean that the stars not only have different temperatures but are made of different materials?

At the Harvard College Observatory, about 1885, astronomers began photographing spectra of stars in large quantity. Instead of putting the slit of a spectrograph at the focus of their telescopes, they made a large, thin prism of glass and fastened it in front of the telescope lens. The light of each star was spread out into a small spectrum, and one photograph showed hundreds of little spectra instead of star images.

The Harvard astronomers decided to classify these spectra. The work was done by Annie Jump Cannon, who published her first classification in 1901 and continued her studies of thousands of spectra until her death in 1941. She labeled these "spectral types" A, B, C, D, E, F, and so on, and made a collection of "standards." There is a great variety of stellar spectra, and they run in a different order: O, B, A, F, G, K, M. This is because the Harvard astronomers later discovered that O and B spectra came from blue stars, F and G spectra from yellow stars, and K and M spectra from red stars. In fact, the star temperatures can be measured (by Planck curves) with the results shown in Table 15. That is, the differences between stellar spectra are mainly due to different *temperatures*, not to differences in compositions.

Type	Appearance	Temperature	Color
O	A few ionized helium lines	30,000°	−0.6 (blue)
B	Weak hydrogen and helium lines	20,000°	−0.3
A	Strong hydrogen lines	10,000°	0.0 (white)
F	Moderate hydrogen and metal lines	7,500°	0.2
G	Strong metal lines	5,700°	0.5 (yellow)
K	Metal lines predominate	4,500°	1.0
M	Molecular bands	3,000°	2.0 (red)

Table 15 Characteristics of standard spectral types of stars.

So the sequence of spectral types, *O, B, A, F, G, K, M,* is a temperature sequence from very hot to fairly cool stars. It is widely used by astronomers who speak of the sun as a "*G*-type star" or just as a "*G* star." They remember the sequence of letters by the ditty, "Oh, Be A Fine Girl, Kiss Me"—an easier memory-jog than Bode's law.

The sequence is continuous; some spectra fall between standards *A* and *F* and are classified *A5.* With a little practice, anyone can classify spectra in eight or ten such classes: *A0, A1, A2,* and so on, and gauge the temperature of a star all the way from 3000° (*M* stars) to 50,000° (*O* stars). Since the steps in temperature are not all equal, if you make an error of $\frac{1}{10}$ of a spectral type for an *M* star, you misjudge the temperature by about 100°K. A similar error between *B0* and *B1* is an error of 2500°K.

Not only the lines of hydrogen, but those of other elements as well must be studied in order to classify the spectra which contain them. By considering all the spectral lines and their meaning in terms of atomic activity and the intensity of the energy in various wavelengths (color), the temperature of each star can be calculated; and it can be assigned to the proper spectral class. Representative types are shown in Table 15.

In 1911, a Danish astronomer, E. Hertzsprung, compared the colors and luminosities of stars within several star clusters. These clusters are so far away

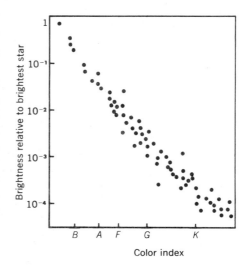

Fig. 20-2 Brightnesses and colors in a typical cluster. Notice that the vertical scale is logarithmic.

that their distances cannot be determined by parallax. However, all of the stars within one cluster are at just about the same distance from us — held together by their gravitational attraction. So the luminosities of the stars in one cluster are proportional to the brightnesses we measure. When Hertzsprung plotted the brightness of each star against its color, studying many clusters but using a separate diagram (like Fig. 20-2) for each cluster, he found that the stars are not distributed all over the plot at random. They do not show all combinations of brightness and color. Instead, most of them lie in a band which slopes from highly luminous blue stars down to red stars of low luminosity. In these far-off clusters, the very luminous stars are the hot ones. With lower temperature the luminosity decreases.

Two years later, a Princeton astronomer, Henry Norris Russell, compared the spectral classes and luminosities of nearby stars — all those for which parallax had then been measured. From each parallax measurement he got a distance R and could determine the luminosity, $L = bR^2$. His results are shown in Figure 20-3, where each dot is a star, plotted accordingly to its luminosity and surface temperature. In Figure 20-3, stars known to be within five parsecs of the earth are indicated by x. Their positions on the diagram are determined by their luminosities in suns and their spectral types.

Russell found that the nearby stars, like those in distant clusters, do not show all combinations of luminosity and temperature. The majority of them lie along a

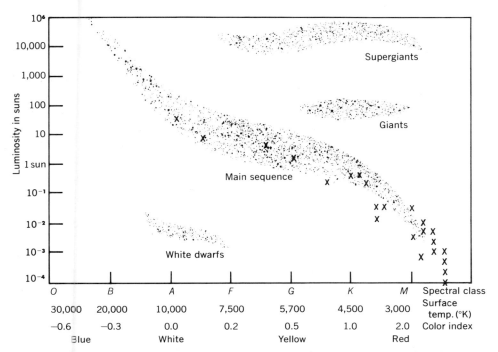

Fig. 20-3 The Hertzsprung-Russell diagram. Each dot is a star; each x is a star within 5 parsecs. The vertical scale is logarithmic.

narrow band which he called the "main sequence." Russell's work showed that the relationship between temperature and luminosity held for the nearby stars. Hertzsprung's work showed that it held for more distant stars. Therefore, diagrams like Figure 20-3 are known as Hertzsprung-Russell, or H-R, diagrams.

Diagrams like Figures 20-2 and 20-3 provided a means of measuring the distances to star clusters. When the main sequence of an H-R diagram of a cluster is laid on top of the one for the nearby stars, so that the two main-sequence bands and the temperatures match, the luminosity of the cluster stars can be read from the left side of Figure 20-3. In each case, of course, it is different from the brightness b used to plot the cluster stars. The difference tells the distance of the cluster in the formula $R^2 = L/b$ (from $b = L/R^2$). For instance, let us say that a cluster contains a star whose spectrum indicates that it is a star like the sun. Since the sun's luminosity is 1, we assume that this star also has luminosity 1. Its brightness, however, is only one millionth that of a zero-magnitude star ($b = 10^{-6}$). Therefore, $R^2 = 1/10^{-6} = 10^6$ and $R = 10^3$ or 1000 parsecs.

You will notice that a number of stars on the H-R diagram (Fig. 20-3) lie above the main sequence, in the upper right of the plot (cool but highly luminous stars). These stars must be larger than those that lie below them, at the same temperature, on the main sequence. These stars are called "giants," and it is easy to see why. For instance, let us consider a giant with a temperature of 3000°K (half the sun's temperature) and a luminosity 10^4 times that of the sun. Stefan's law (p. 96) tells us that $E \propto T^4$. Therefore, each square centimeter of the stars emits only $(1/2)^4$ or $1/16$ as much light as each square centimeter of the sun. So the surface area of the star must be $(16)(10^4)$ times that of the sun, or 160,000 times larger than the sun's. This means that its radius is 400 times the sun's radius. If the center of our solar system were placed at the center of such a star, the star's surface would lie beyond the orbit of Mars.

A normal red star that lies below this giant on the main sequence (10 million times less luminous) also has a surface temperature of 3000°K. Its luminosity, however, is $1/1000$ that of the sun. As in the case of the giant, each unit area of this star emits only $1/16$ as much light as a unit on the sun's surface. But, in order to have the calculated luminosity, its surface area need be only $(16)(1/1000)$, or $1/63$, the sun's surface area, and about $1/8$ the sun's radius. Its size is not too different from that of the sun.

Above the giants on the H-R diagram are stars of even higher luminosity and almost as low surface temperatures. These are called "supergiants" and they have even larger radii than giants.

There are, in addition, stars in the lower left-hand corner of the H-R diagram (hot, with low luminosity) which are known as "white dwarfs." In spite of their low luminosities, they have very high surface temperatures, so they must be very small—as stars go. A typical one has a surface temperature twice that of the sun, yet its luminosity is $1/200$ that of the sun. Its surface can be only $1/(200)(2^4)$, or $1/3200$, of the sun's area, and its radius $1/57$ that of the sun.

So there are stars of many different sizes as well as many different temperatures. Although we started out to show that the sun is like the stars, we have ended the

154

chapter by concluding that the stars differ greatly among themselves. Yet the sun does appear to be one of them, of middling size and temperature. It differs from the other stars no more than they differ from each other. And there are many stars (such as Alpha Centauri) whose only differences from the sun appear to be their distance from us.

Additional Reading

> McCREA, W. H., "Stars: Data and Classification" in *Astronomy* (Samuel Rapport and Helen Wright, eds.): New York, New York University Press, 1964.

chapter 21 | The Inconstant Stars

Back in Ptolemy's day, anyone who wanted to be an army officer had to have good enough eyesight to see that there are two stars at the bend of the Big Dipper's handle. About 11′ away from the brighter star, Mizar, is another star, called Alcor, only one-fifth as bright. In 1650 an Italian astronomer, John Riccioli, took a good look at Mizar through his telescope and was amazed to see that it is not just one star. He saw a pair of stars much like Mizar and Alcor but only 14″ apart. As more and more astronomers began to use telescopes, many more stars were seen to be double. In most of them one star was considerably fainter than the other.

Because the two stars in a pair are so close together in the sky, it is evident that they lie in almost the same direction from us. They could be stars with the same luminosity at different distances or they could be stars of different luminosity at the same distance. Sir William Herschel and his sister Caroline, who found over 700 double stars with their telescope, believed that the fainter star of each pair was far more distant. Because it is easy to measure a small angle between two stars in a pair, they planned to obtain the parallax shift (p. 64) of the nearer (brighter) star in each pair by accurately measuring the angle between it and its far distant companion at six-month intervals. In this way, the distances of many stars could be measured. If the two members of a double-star system were at greatly differing distances, each year as the earth travels its orbit, parallax would make the nearer one shift back and forth. (At this time, 1782, no stellar parallax had been observed.) To their disappointment, the Herschels failed to find any yearly back-and-forth motion.

Instead of parallax, however, they discovered another sort of motion. In 1804, while comparing new measurements of the double star Castor (Fig. 2-6c) with some made 22 years earlier, Herschel found that the bright star and its fainter companion were in slightly different positions relative to each other. (Figure 21-1

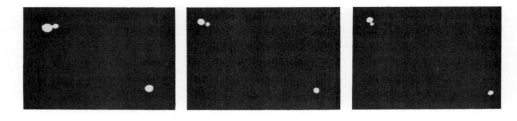

Fig. 21-1 Photographs of a double star, Krueger 60, made in 1908, 1915, and 1920, showing the orbital motions of the two stars around each other. [Yerkes Observatory photograph.]

shows similar changes in another double star.) This could only mean that the two stars which make up Castor are moving around each other, like the moon around the earth. Herschel could not tell whether just one or both were in motion, because his measurements were only of the angular distance between them and the direction of one from the other. Nevertheless, it looked as though they were moving around their center of mass. Newton had said that his laws are universal, but this was the first time that they could be confirmed outside the solar system.

Multiple systems like this are common. There are also much larger groupings of stars, called clusters. Most of the stars of the Big Dipper, together with many stars nearby in the sky, form a cluster. These, as we shall see later, have yielded a wealth of information about the lives of the stars.

About every three days, the bright star Algol in the constellation Perseus (Fig. 2-6d) fades to a third of its usual brightness, a change which can be observed without a telescope. After a few hours it returns to its normal brightness. Its changing brightness was probably known to the Arabs as long ago as A.D. 900, for they named the star "Al Ghūl," which meant "changing spirit." However, it appears not to have been noticed by the Greek or European astronomers until at least 1669, which is surprising.

In 1783, Algol's varying brightness was first measured by John Goodricke, an 18-year-old English boy, deaf and mute. He found that the bright-to-dim cycle repeated itself every 69 hours (Fig. 21-2a) and suggested that the cycle might be due to large, dark spots on the star (like sunspots) which are turned toward us for a few hours during each rotation of the star. However, he believed it more likely that Algol is made up of two stars, one of low luminosity revolving around another of high luminosity, in an orbit edge-on to the earth (Fig. 21-2b). Once every 69 hours, the dimmer star passes in front of the brighter star, cutting off its light. Midway between these eclipses, the brighter star, in turn, passes in front of the dimmer one. This accounts for the small dip in the light curve that is shown in Figure 21-2a.

At the time that Goodricke made this suggestion, most astronomers thought, as Herschel did (p. 155), that double stars merely happened to be near each other on the celestial sphere and were far from each other in space. Furthermore Herschel's telescope, the best in England, failed to show Algol as two stars. Good-

156

ricke died at 21, 20 years before Herschel himself found that double stars do indeed revolve around each other, and a hundred years before the shifting lines in Algol's spectrum were measured. These lines prove that Algol revolves in an orbit and indicate that it has a companion star too dim to be seen or recorded in the spectrum.

These "eclipsing binaries" are important because their masses can be accurately determined by means of Kepler's and Newton's laws. The mass of Jupiter can be calculated from the motions of its moons and the mass of the sun by the orbital motions of its planets. In much the same way, the masses of double stars, moving around their centers of gravity, can be determined. However, Newton found that when the masses are nearly the same (as in the case of double stars), both must be used in the calculations, and the formula $M_1 = R^3/p^2$ must be written $M_1 + M_2 = R^3/p^2$. Because the orbits of an eclipsing binary are edge-on to us, it is easy to work out the mass, and such stars are searched for eagerly.

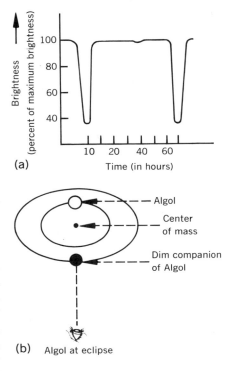

(a)

(b) Algol at eclipse

Fig. 21-2 (a) The light curve of Algol. (b) Algol and its unseen companion at the time of each sharp dip in the light curve of (a).

Astronomers have found the masses of about 50 of the millions of stars in the sky. Of course, these are all double or multiple stars. However, their masses are believed to be typical because a large proportion of the stars are double or multiple, and almost all spectral types and luminosities are represented.

The surprising thing is that the range of stellar masses is so small—from one-tenth the sun's mass to 75 suns. (Luminosity of the stars, you recall, varies from $1/100$ to 100,000 times that of the sun.) It is not surprising to find that the more massive stars are more luminous. Mass, however, is not directly proportional to luminosity; a star twice as massive as the sun is about 11 times as luminous. When mass is plotted against luminosity, as in Figure 21-3, most of the stars fall along a narrow sequence running from the upper left corner of the diagram (high mass, high luminosity) to the lower right (low mass, low luminosity).

All of the stars on the main sequence of the temperature-luminosity (H-R) diagram (Fig 20-3) fall on this mass-luminosity sequence. The amount of material that a main-sequence star contains (its mass) seems to determine how luminous it is. This suggests that stars on the main sequence are all made up of much the same material (have similar chemical compositions), since the *amount* of material seems to determine how much radiant energy the star emits. We might also guess

157

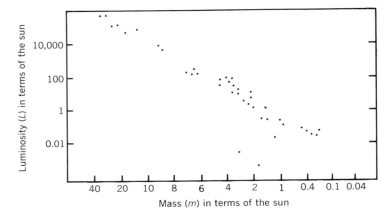

Fig. 21-3 The mass-luminosity relation for the fifty stars — all members of double-star systems — whose masses have been determined by their orbital motions. The sun, a member of a star-planet system, is also included; it falls, of course, at 1 on the horizontal scale and 1 on the vertical scale. Both the vertical and the horizontal scales are logarithmic.

that the *structure* of the star (the way the material is arranged) would be similar, and that the nuclear processes which heat these main-sequence stars are similar. Since our sun is one of the stars on this sequence, it seems probable that its energy source, the conversion of hydrogen to helium, is also the energy source of the main-sequence stars.

By somewhat more complicated calculations, we can determine the diameters of these stars, as well as their masses. Knowing the diameters and volumes, we can calculate the densities. The densities of stars on the main sequence range from about 0.02 to 10 times that of the sun, which again indicates that their composition and structure are not too different from each other and from the sun; they are all large masses of hot, glowing gas.

You will recall that the distances of main sequence stars, which are too far away for parallax measurements to be made, can be determined from their brightness and their position on the H-R diagram (p. 154). In a similar way, the masses of double stars which cannot be determined by observation, as well as the masses of single stars, can be determined from the mass-luminosity diagram. A star whose spectrum shows that it belongs on the main sequence of the H-R diagram can be fitted into its proper place on the main sequence, according to its temperature (color), and its luminosity can then be read from the diagram. Then it can be placed on the mass-luminosity diagram according to this luminosity and its mass read directly from that diagram. Astronomers have confidence in this method because the masses of double stars that can be determined from observation agree with the masses arrived at by fitting the star into the H-R and mass-luminosity diagrams on the basis of spectral type and brightness alone.

As an example, let us consider a star whose spectrum shows its surface temperature to be 10,000°K. It is of spectral type *A*, and the lines in its spectrum indicate that it is not a white dwarf. The H-R diagram shows that its luminosity is

40 times that of the sun. On the mass-luminosity diagram, this luminosity indicates that the star's mass is 3.5 times that of the sun.

If the star had the same surface area as the sun, its temperature (1.4 times the sun's) would give it a luminosity about four times that of the sun, for according to Stefan's law, $E \propto T^4$ and $(1.4)^4 = 4$. But the star's luminosity is 40 times that of the sun, so its surface area must be 10 times that of the sun. Its radius r, therefore, is $\sqrt{10}$ or about 3.2 times that of the sun; and its volume r^3 is 31.5 times that of the sun. The star's density (mass/volume) is $^{3.5}/_{31.5}$, or 0.11, times that of the sun.

In this same way, many star masses and densities have been determined, making it possible for astronomers to learn much more about the universe of stars. Much of this knowledge would never have been gained without the measurements of double stars, which established the mass-luminosity relation.

A year after John Goodricke showed that Algol's changes in brightness are regular in amount and period, he found another changeable star, Delta (δ), in the constellation Cepheus (Fig. 1-2). As he had done for Algol, Goodricke drew the light curve of this star (Fig. 21-4) by putting together measurements made on many different nights. (Neither star can be observed continuously through a complete cycle because daylight interrupts the observations.) Like Algol, δ Cephei's brightness decreases by a third, and its period ($5\frac{1}{2}$ days) is not too different from Algol's. However, the shape of Algol's light curve (Fig. 21-2a) is different. δ Cephei rises rather rapidly to maximum light and then falls more slowly to minimum light, while Algol just has brief dim intervals (during eclipse).

Goodricke did not suggest that δ Cephei is an eclipsing binary like Algol. However, his explanation of Algol's varying brightness was so well accepted, after Herschel found so many double stars, that astronomers thought all stars with periodic changes in brightness (including δ Cephei) must be eclipsing binaries. This view persisted for almost a century.

Meanwhile, hundreds of "cepheids," stars with light curves similar to that of δ Cephei, were found. Among them is Polaris whose brightness varies by 10% in a period of about four days. Gradually the evidence indicated that cepheids cannot be eclipsing binaries and that their brightness changes have a different cause.

It is hard to understand today how people could think that it was possible for an eclipse to give the light curve of δ Cephei, for its shape is wrong. An eclipsing star moves onto the other star at the same velocity with which it moves off. This makes the dips in the light curve symmetrical (the same shape dimming as brightening). This is not the case in a cepheid's light curve.

Doppler shifts are observed in the light

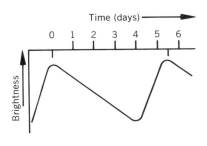

Fig. 21-4

from a cepheid. However, the spectra contain only single lines at the various wavelengths. Of course, this does not mean that a second star is not present. It could be too dim to produce lines in the spectrum which can be seen in a photograph. If this were the case, the eclipsing binary would be dimmest at the time when the brighter star is eclipsed — when its velocity toward us is zero. The velocity curve of a cepheid, however, shows that it is dimmest when the star's motion toward or away from us is greatest.

The spectral type of a cepheid changes. δ Cephei at maximum brightness is an *F6* star; at minimum light it is a *G1* star. Both spectra are equally distinct, showing that if there were two stars, one could not be much dimmer than the other. But then there ought to be double lines, caused by opposing Doppler shifts, at those wavelengths where both stars have absorption lines. Furthermore, if there were two stars, one of type *F6* and the other of type *G1*, their spectra would be mixed at maximum light. When neither star is eclipsed, lines characteristic of both types should be present in the spectrum; but this does not happen. In all these ways, the spectra of cepheids are hard to explain as those of eclipsing binary stars.

Here was a puzzle. It looked as though each cepheid is a single star which varies in brightness, spectral type, and temperature. At the same time, Doppler shifts indicate that the star continually repeats a cycle in which it first approaches us, then stops and recedes, then stops and approaches us again. Newton's laws tell us that a star cannot move like this unless there is another star nearby. And there seems to be no other star. Many different explanations were suggested, but none seemed to really explain the cepheids.

Then in 1914, Harlow Shapley, a young astronomer at Mount Wilson Observatory in California, recalled a suggestion made in 1879 when little was known about the cepheids except their light curves and temperature changes. Almost ignored and all but forgotten, it explained the cepheid's changes in luminosity as due to temperature changes, caused in turn by expansions and contractions of the star. According to this theory, a cepheid beats like a giant heart. Shapley pointed out how well this explanation fits the information from Doppler shifts. The star as a whole does not move back and forth; instead, its photosphere (the surface from which its light is coming) first moves outward and then inward, toward us and away from us, causing the changes in Doppler shift of the spectral lines. Instead of the whole star moving around an orbit so as to recede, approach, then recede, it is just the half of the star's photosphere facing us which moves at changing velocity, first toward us, then away (Fig. 21-5). The changing velocities calculated from Doppler shifts are the true motions of the photosphere on our side of the star, first outward, then inward.

But what could make a star expand and contract like this? A few years after Shapley proposed this pattern, the English scientist, Sir Arthur Eddington, explained it by the known laws of physics. He pointed out that in a normal star, like the sun, the pull of gravity toward the center of the star is balanced everywhere by the outward pressure of the gas molecules and by radiation pressure. This keeps a normal star at one "balanced" size. If a star is compressed somehow (like a soft rubber ball is squeezed), the gravitational force pulling each gas molecule

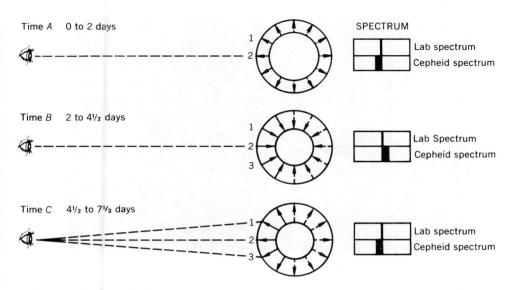

Fig. 21-5 Movements of the photosphere of a cepheid toward and away from an observer on earth, as shown by spectra. The spectral lines are shifted toward the violet at *A* and *C* and toward the red at *B*. The lines are wide because points 1 and 3 do not have as large a motion in our line of sight as does point 2.

toward the center becomes greater because the molecule's distance from the center is less. However, laboratory experiments show that when a gas is compressed into a smaller space, its temperature and pressure increase. (Heating the gas increases the motion of its molecules and this increases the pressure.)

Eddington's calculations showed that this outward gas pressure would be somewhat greater than the increased inward force of gravity. Thus the star would expand and, in time, return to its original size. However, it would overshoot its "balanced" size, due to Galileo's principle of inertia (p. 56), which indicates that a force is needed to stop the molecules outward motion. The inward pull of gravity (and collisions with other molecules) would slow down the outward motion but the molecules would not stop instantly. So when the star reached its balanced size it would still be expanding, and before the gases stopped moving outward, the star would be larger than normal. (We are observing the same sort of phenomenon that is shown by a pendulum. The bob of the pendulum does not stop at the bottom of its swing but continues on past even though there is a force acting to retard this motion.) When the star overshot its "balanced" size, the outer layers would not be fully supported by the gas pressure below and the star would collapse. This contraction starts the cycle again. Once started, then, the pulsation would continue, like the swinging pendulum or a bouncing ball. But how it got started is still unknown. As you may have guessed, it could be started by an explosion deep in the star. In this case, the cycle would begin with an expansion of the star.

Fig. 21-6 The Large (left) and Small (right) Magellanic Clouds, photographed at Harvard's South African Station. These systems of stars are too far south on the celestial sphere to be visible from the United States. [Harvard College Observatory photograph, courtesy of Yerkes Observatory.]

From his mathematical studies, Eddington predicted three conditions which must be met if cepheid brightness changes are actually due to pulsations. (1) The increase from the balanced size should be less than 10%, as should the decrease. As we have seen, for δ Cephei it is 6%, and the amounts for other cepheids are all less than 10%. (2) The temperature in the interior of the star would be highest when the diameter is smallest (when the pressure is greatest). However, the heat will take some time to work out to the surface. Therefore, the greatest luminosity will come some time later, while the star is expanding rapidly. Similarly, the minimum temperature will come while the star is contracting. This is, indeed, the case for δ Cephei; it is so for other cepheids as well. (3) The period should be inversely proportional to the square root of the star's density ($P \propto \sqrt{1/D}$ or $P = \sqrt{k/D}$).

In 1912, Henrietta Leavitt at the Harvard College Observatory had studied many photographs of two large groups of stars called the Magellanic Clouds. In the Small Cloud (Fig. 21-6), Miss Leavitt had identified 25 cepheids and plotted their light curves. She found that the periods of these stars are related to their relative brightnesses, as shown in Figure 21-7. The brighter ones all have longer periods. A cepheid with a period of 30 days, for instance, is six times brighter than one whose period is three days. Since these stars are all in the same star group, they are all at about the same distance from us; and so their relative brightnesses are the same as their relative luminosities. Since the distance of the Cloud was not known, the actual luminosities of these cepheids (compared to the sun or some other standard) were unknown.

Shapley saw that Miss Leavitt's discovery was evidence for the pulsation theory. A more luminous star must be bigger than a less luminous one of about the same spectral type (and most cepheids are either F or G stars). The larger a star is, the longer the pulsation process should take (for the same reason that a large organ pipe gives a low-pitched note).

He saw also that the discovery was important for another reason. If the distance R of the Magellanic cepheids could be determined, then the true luminosity L of each could be calculated, since their brightness b could be measured, and $L = b/R^2$. Each Magellanic Cloud also contains many ordinary stars and the spectra of these stars show that they lie on the main sequence of the H-R diagram. So, unlike the cepheids, their luminositites could be read directly from the H-R

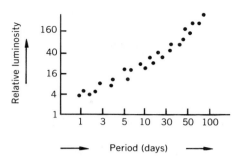

Fig. 21-7 The relative brightness (or luminosities) of the cepheids in the Small Magellanic Cloud.

diagram and their distances determined from $R = \sqrt{L/b}$. This distance is, of course, the same as that of the cepheids in the Clouds. Now, knowing both b and R, Shapley could determine the luminosity L (compared to the sun), of each Magellanic cepheid and replace the relative luminosities (Fig. 21-7) with the true luminosities. There was no reason to suppose that these cepheids were different from other ones. Now to get the distance of other isolated cepheids, he had only to compare them to the true luminosities of the magellanic cepheids (according to their periods), determine their true luminosities, and calculate $R = \sqrt{L/b}$. The distances of the isolated cepheids range from 60 to 6000 parsecs. In addition, the distance of any star cluster or star cloud that contains cepheid variables can be determined. As you may have suspected, cepheids are very luminous stars — many of them 10,000 times the sun's luminosity — and so can be seen more easily than other stars in such a group.

Knowing both the spectral type and the luminosity of many cepheids, Shapley could fit them into the H-R diagram (Fig. 20-3). They are the thin scattering of stars above the main sequence and to the left of the red giants — more luminous than main-sequence stars of the same spectral type and of higher temperature than red giants of the same luminosity.

Unlike white dwarfs, red giant stars fit on the mass-luminosity sequence (Fig. 21-3) in the cases where masses are determined from double star motions. Therefore, it seemed reasonable that cepheids, similar to the giants spectral type and luminosity, might also fit. If they did, then masses could be read from the M-L diagram. At least, it was worth a try. When their diameters were also calculated, their densities could be obtained. Densities of a few of these stars are listed in Table 16. How well do they fit Eddington's prediction, $P = \sqrt{k/D}$? This equation can, of course, be written $k = P\sqrt{D}$. Notice that the values for $P\sqrt{D}$ in the right-hand column of Table 16 are all nearly the same — k does indeed appear to be a constant. Eddington's third requirment is fulfilled, and it looks as though the

Star	Period P (in days)	Density D (sun = 1)	$\sqrt{D}$	$P\sqrt{D}$
SV Cassiopeiae	1.95	0.0030	0.050	0.10
α Ursae Minoris	3.97	0.0008	0.028	0.11
δ Cephei	5.40	0.0006	0.024	0.13
η Aquilae	7.18	0.0003	0.017	0.12
ζ Geminorum	10.15	0.0002	0.014	0.14
X Cygni	16.38	0.00008	0.009	0.15
Y Ophiuchi	17.12	0.00008	0.009	0.15
I Carinae	35.52	0.00003	0.005	0.18

Table 16 Period-density information on characteristic cepheids.

cepheids are pulsating stars. (It also means that Shapley's hunch that they fall on the M-L sequence was correct.)

Cepheids are *variable* stars; their periodic changes in brightness are due to changes within the star itself. They are not the only sort of variable star. In Figure 21-8 the types which differ from each other in period, luminosity, luminosity change, and spectral type are placed on the H-R diagram. All of them are probably pulsating stars. In addition, there are variable stars whose brightenings do not occur on schedule and whose light curves are not always the same shape. These unpredictable light variations may be due to "starspots," suggested by Goodricke as a possible source of Algol's light variations (p. 156), or they may be pulsating stars in eclipsing binary systems.

At least some of the stars, then, are not perfect and unchanging, as the followers of Aristotle thought. Tycho Brahe, you will recall, began to have some doubts

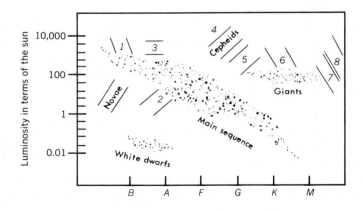

Fig. 21-8 The Hertzsprung-Russell diagram for types of variable stars. Notice that the cooler stars have longer periods. See table 17.

Number in figure	Type	Period	Spectrum	Luminosity (suns)	Luminosity changes (suns)
1	β Canis Majoris	4 hours	B	1600	1400 to 1800
2	Dwarf cepheid	3 hours	A and F	16	10 to 25
3	RR Lyrae	12 hours	A and F	100	60 to 160
4	Cepheid	7 days	F and G	1600	1000 to 2500
5	W Virginis	15 days	F and G	630	400 to 1000
6	RV Tauri	75 days	G and K	630	250 to 1600
7	Short long-period	175 days	M	250	8 to 800
	Long long-period	350 days	M	100	6 to 1600
8	Semi-regular	100 days	M	630	400 to 1000

Table 17 Characteristic periods, spectra, and luminosities of representative stars from Figure 21-8.

about Aristotle's ideas when he saw a nova or "new star" in the constellation Cassiopeia. He thought it really was new, but telescopic photographs of other novae show that these "guest stars" (as the Chinese call them) were already there. They are ordinary faint stars which suddenly become as much as several thousand times more luminous. Some of them, not too far away, become visible as bright stars in the sky. They stay bright for a few days or weeks, then gradually fade to their former luminosity and become ordinary stars again.

What happens as the star suddenly brightens and slowly fades? Its changing spectrum and photographs like Figure 21-9, taken after the nova had faded to normal brightness, show that an outer layer of the star is ejected as a shell of gas. This rapidly growing shell first gives the effect of an expanding photosphere and accounts for most of the increase in luminosity. As the shell of gas moves outward, its volume increases and its density drops. Then it becomes transparent, but it still glows because its atoms absorb the star's ultraviolet light and emit the energy as visible emission lines. As the shell continues to move outward, it gets farther from the star. Then each square mile does not receive as much light and so cannot emit as much. It

Fig. 21-9 Photograph made with the world's largest reflecting telescope (200-inch working diameter) of the shell of the nova in the constellation Perseus, which brightened in 1901. The picture was made almost 60 years after the nova's outburst. The star, now back to normal size, is clearly visible at the center of the nebula (shell). [Mount Wilson and Palomar Observatories.]

165

becomes a large, dim *nebula,* or cloud of gas (Fig. 21-9), rather than a smaller bright source of light. The star itself looks like it did before the outburst; the only difference is that it has lost a small amount of mass. (A nova in the constellation Hercules, that brightened spectacularly in 1934, is a member of a double-star system. Since then, the period of the system has increased by 3½ minutes, indicating a small decrease in mass.)

Certain very rare novae, appropriately called supernovae, light up to several hundred million times their former brightness. Tycho's star of 1572 was one of these, as well as one which Chinese astronomers observed in A.D. 1054, in the constellation Taurus (Fig. 2-6c).

Additional Reading

ABELL, G. O., *Exploration of the Universe:* New York, Holt, Rinehart and Winston, 1964, Chaps. 21 and 27.

CAMPBELL, L., and L. JACCIA, *The Story of Variable Stars:* New York, McGraw-Hill, 1941.

PAGE, THORNTON, and L. W. PAGE, eds., *Starlight:* New York, The Macmillan Company, 1967, Chap. 7.

chapter 22 | **Lives of the Stars**

After a star has expelled a shell of gas, it is not quite the same as it was before it became a nova. Its mass is slightly smaller—much smaller if it had been a super-nova. These changes in a few stars are spectacular and quick. The rest of the stars seem to be the same, year after year; but even in them slower changes must be going on. Early astronomers thought of the stars as eternal and unchanging, like jewels decorating the celestial sphere. When modern astronomers realized that the stars are pouring out energy at a tremendous rate, it became clear that each star must be changing and cannot shine forever. The stars are using up their own material as they convert mass into energy and, as hydrogen is changed into helium in its core, a star's chemical composition is altered.

So the stars should be slowly changing, or ageing, but there seems to be no evidence that they are doing so. The bright stars listed over 1800 years ago in Ptolemy's *Almagest* can still be seen as bright stars. Reddish stars that Tycho Brahe described in the sixteenth century are still reddish in color. The cepheids have periods of length expected for their densities, and no steady changes in period have been observed, even though timing the pulsations over 10 years can determine the period of a cepheid to one part in 10,000. Slight changes measured in the periods of one or two cepheids reversed themselves — disappeared — after five years. No steady change seems to have taken place while men have been looking at the stars and recording what they saw. Perhaps this only means that the rate of change is very slow.

Nevertheless, if the stars do change steadily, one result ought to be visible: Unless all the stars formed at the same time and changed at the same rate, there should now be stars in various stages of development. As we have seen, there are different types of stars: red giants, main-sequence stars, white dwarfs, and variable stars (Fig. 21-8). Each type may represent a different stage in the life of a star, like childhood, youth, and middle age. Or perhaps they do not; maybe each different type was formed the way it is.

Most stars belong to the main-sequence group. Next in number (per unit volume) are the white dwarfs, while the red giants and other types are much rarer. Within the main-sequence group, K- and M-type stars of low mass and low luminosity are most numerous. Are these small, faint, main-sequence stars common because more of them were produced? Or are there more of them because a star stays longest in this stage of its life? This may be like asking why there are more adults than children.

There is one way to decide whether the different types of stars we see are stages in the life of any star. Since we know the structure and composition of the stars, as well as how they produce their energy, a theory of stellar evolution can be worked out, using the laws of nuclear physics and of how gases behave. Then the star types observed in the sky can be compared with the stages of development predicted by this theory. Do they fit in? Would a star be expected to shine steadily for a long time in its main-sequence stage? Which type of star is oldest? Which is youngest? Early in this century, astronomers began thinking about these questions. The theory of stellar evolution is still far from complete, and many of its details may change before the century is over.

As we have seen, the sun appears to be a sphere of gas held together by gravitational attraction. Nuclear reactions convert hydrogen to helium in the sun's central core, where the pressure of overlying material keeps the density and temperature high. The energy produced by the nuclear reactions in the core works its way through the outer layers of the sun, continually being absorbed and re-emitted, until it reaches the photosphere. Then we see it as sunlight.

When the hydrogen in the core is all converted to helium, no more energy can be produced by that nuclear reaction there. Then the sun will no longer be a main-sequence star. From estimates of the amount of hydrogen that it contains and from measurements of its luminosity (the rate at which it is producing energy), astronomers calculate that the sun can remain a main-sequence star for about 15 billion (1.5×10^{10}) years more. Evidence from earth's oldest rocks indicates that the sun was shining with about the same luminosity 4 billion years ago. Thus, the total time it will spend in the main-sequence stage (I in Fig. 22-1) may be about 2×10^{10} years.

The M-L diagram shows that the luminosity and surface temperature of a star on the main sequence depend only on its mass. This suggests that in all other ways

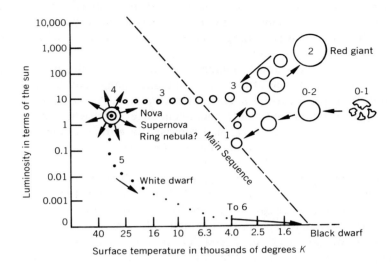

Fig. 22-1 The evolution of a *G2* star (like the sun) shown on the H-R diagram. Arrows do not indicate movement of the star; they show the order of its stages of evolution, as predicted by present-day theory.

(chemical composition, structure, and nuclear processes) the main-sequence stars are similar. You might think that the more massive ones would remain in the main-sequence stage longer since they have more hydrogen to spend. However, the M-L diagram shows that more massive stars are much more luminous. Then you might guess that all stars would remain in the main-sequence stage for the same length of time. And they would, if mass were directly proportional to luminosity. The M-L diagram shows, however, that $m \propto L^{3.5}$. A star twice as massive as the sun is not twice as luminous, but instead $2^{3.5}$, or 11.3, times as luminous. It is using up the energy in each gram of its mass $11.3/2$, or 5.65, times as fast, and can remain in the main-sequence stage only $1/5.65$ as long as the sun.

The time T that a star will stay in stage *1* (Fig. 22-1) is thus given by the formula $T \propto m/L \propto m/m^{3.5} \propto m^{-2.5}$. Or the formula can be written $T = T_S/m^{2.5}$ (where T_S is the life of the sun on the main sequence and m is the star's mass in terms of the sun's mass). Stars of low surface temperature (and therefore low mass) remain in the main-sequence stage longer (Table 18 and Fig. 22-2). These stars (of spectral types *G*, *K*, and *M*) are the most common main-sequence stars. The suggestion is strong that they are most common because they last longer than the others, and accumulate like bottle caps in a trash burner.

Of the hydrogen used up, only 0.7% of the mass is converted to energy; the star does not change its mass very much during its life in the main-sequence stage. However, the chemical composition of the central part of the star gradually changes from mostly hydrogen to mostly helium. Finally there comes a time when nuclear energy can no longer be produced by converting hydrogen to helium be-

Spectral type	Mass (1 = sun's mass)	Luminosity (1 = sun's L)	Expected time on the main sequence (years)
O	80	280,000	6×10^6
B	4	2,000	4×10^7
A	2.5	52	10^9
F	1.5	5	6×10^9
G	1	1	2×10^{10}
K	0.8	0.3	5×10^{10}
M	0.15	0.01	3×10^{11}

Table 18 Lengths of time that the various spectral types are expected to remain on the main sequence.

cause the central part of the star (where conditions permit nuclear reactions) is all helium. The main-sequence stage (Fig. 22-1, stage *1*) is over.

But the life of the star is not over. As the hydrogen nuclear processes slow down, energy is generated by another process, and in stage *2* (Fig. 22-1) it takes

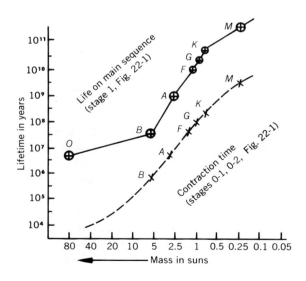

Fig. 22-2 The upper curve shows the length of time that various types of stars can remain in the main-sequence stage (stage *1* of Fig. 22-1), plotted against the star's mass. The data is from Table 18. The lower curve shows the times that each of these stars will take to develop from stage *0-1* to a main-sequence star.

over. As the helium core begins to get cooler, outward movements of the gas and radiation pressure can no longer balance the pull of gravity toward the center. The helium core shrinks, and energy is released by this contraction. (You will recall, from p. 146, that contraction was once thought to be the only source of the sun's energy.) During this contraction, the helium core becomes much hotter than in stage *1*, causing large amounts of energy to flow outward through the star. This causes a rearrangement of its outer layers; they expand and the star grows much larger. Like all expanding gases, these outer layers cool off. The yellow *G*-type star becomes red and large. Although more energy is being radiated by the star as a whole, each square mile of its enlarged surface area is sending out less energy. The star is no longer a main-sequence star; it has become a cool but luminous *red giant*.

As contraction of the core continues, the center of the star grows even hotter. The pressure, temperature, and density now permit the helium itself to change by new nuclear reactions into heavy elements. The energy produced in this way stops further contraction of the core and causes the gas there to expand again, although the outer layers of the red giant shrink to nearly their former size. The new source of energy from helium nuclear reactions keeps the star as luminous as before, but its surface temperature rises because the star's photosphere is now smaller. The star's luminosity and surface temperature now place it at position *3* on the H-R diagram (Fig. 22-1). It is no longer a red giant.

Calculations show that the red-giant stage (Fig. 22-1, stage *2*) would be much shorter than the main-sequence stage (Table 19). A star which had been like the

Stage		Star of mass 4 suns	Star of 1 solar mass	Percentage of the stars within 100 parsecs of the sun that are in each stage
0-1, 0-2	Contraction from a gas cloud	7×10^5 years	7×10^7 years	Negligible
1	Life as main-sequence star	5×10^7 years (as *B2* star)	2×10^{10} years (as *G2* star)	89
2	Red giant stage	10^6 years?	10^8 years?	1
3	Unstable variable stage	10^6 years?	10^6 years?	Negligible
4	Nova stage Ring nebula stage?	10 years? 20,000 years?	10 years? 20,000 years?	Negligible
5	White dwarf stage	10^9 years	10^9 years	10
	Totals	1.1×10^9 years	2.1×10^{10} years	
6	Invisible cold stage	Infinity (the rest of time)	Infinity (the rest of time)	0?

Table 19 Duration of the various stages in two stars' lives.

sun would remain a red giant for about 10^8 years; more massive stars for shorter times (10^6 years, for instance, for a *BO* star). So there are fewer red giants than main-sequence stars in the sky.

What happens during the remainder of a star's life is more uncertain. Many astronomers are busy today trying to figure it out. If you compare Figure 22-1 with Figure 21-8, you will see that stars of about the luminosity and temperature predicted for stage *3* are variable stars of the RV-Tauri type (variable type *3*). Therefore, many astronomers believe that at this stage (which may last for 10^6 years), the star becomes unstable, explosions may take place within it, and it may begin to pulsate. Stars of greater mass than the one whose life history is shown in Figure 22-1 may have different luminosities and temperatures at this stage and may become other types of variable stars.

In stage *3*, helium is turning into heavier elements. These nuclear changes cause the star to become smaller and more dense. The luminosity (energy radiated per second) remains the same, but the star's surface temperature continues to increase because this energy is radiated from a smaller and smaller photosphere. In time, the star's luminosity and temperature become like those of a nova just before it blows up. It is possible that as the star enters stage *4* (Fig. 22-1), it blows off its outer layers. If it is a supernova, it blows off all the outer layers, leaving only its core to remain as a star. It becomes a very dense *white dwarf*, with the luminosity and temperature of position *5* (Fig. 22-1), like the star at the center of the Crab Nebula (Fig. 22-3). Novae, on the other hand, do not become white dwarfs; they are about the same before and after their explosions—stars with the luminosity and temperature of position *4*. Perhaps many nova explosions must take place in a star, each one blowing off some of its outer layers, before the big, final supernova explosion leaves only the dense core, making the star a white dwarf. (Or maybe supernova explosions only take place in stars that are very massive. Astronomers are still arguing this point.) Planetary nebulae (sometimes called ring nebulae) may also be formed at this stage from stars with very hot cores.

By stages *3* and *4* (variable and nova stages), the only thing left to supply energy is the conversion of helium to heavier elements. When a star reaches the white-dwarf stage, all that is left is its small, dense core. In that old core, no more hydrogen or helium remains to produce energy by nuclear change. White-dwarf stars are extremely dense (140,000 times the density of water). They are not spheres of solid material; no known solid material is that heavy. Some white dwarfs

Fig. 22-3 The Crab Nebula. [Mount Wilson and Palomar Observatories.]

are almost a million times the density of the sun. The only material that could be that dense would be electrons and the nuclei of atoms, all pressed tightly together. A large part of the normal atom, such as we have considered up to now, is empty space — space where the electrons orbit the nucleus. No empty space remains in the material of a white dwarf, where nuclei and electrons are jammed together.

This material cannot produce any more energy. Slowly the old core, now a white dwarf, cools off like a poker taken out of the fire. The white-dwarf stage, like the main-sequence stage, is one where stars linger for a long time; perhaps as long as 10^9 years. So we should find many white dwarfs. They make up only 10% of the stars within about 100 parsecs of the sun. Even so near, their low luminosity makes them hard to find. The more distant ones are impossible to see.

Calculations show that in a few hundred million years, a white dwarf fades to 1% of the sun's luminosity; in several billion years, to $1/10,000$. In time it will cease to shine at all and become a "black dwarf" (Fig. 22-1, stage 6) — a cold, non-luminous sphere of dense gas. It would take a white dwarf many trillions of years to reach this stage. It is possible that there are no black dwarfs among the stars because the universe may not be this old.

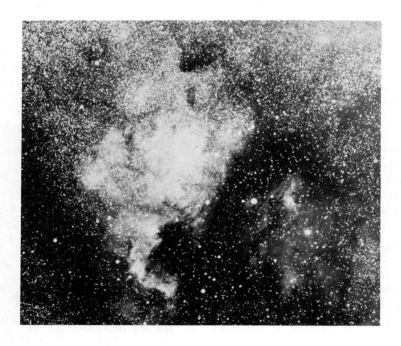

Fig. 22-4 The "North-America" nebula in the constellation Cygnus, photographed with a 10-inch reflecting telescope. [Yerkes Observatory photograph.]

The upper curve of Figure 22-2 shows that if *O, B,* and *A* stars had become main-sequence stars at the same time as the sun did (4×10^9 years ago), they would no longer be main-sequence stars. By now they would all be red giants. But there are lots of *A* and *B* stars, and quite a few *O* stars in the sky. If the theory of evolution of main-sequence stars is correct, this means that the stars were not all born at the same time. It also suggests that stars are being formed right now.

What would a "new-born star" look like? Do stars suddenly begin their lives on the main sequence, as "balanced" spheres of gas with nuclear processes going on efficiently in their cores? Astronomers do not think so. Main-sequence stars seem too mature to be brand new. Some process has to get gaseous material together in individual spheres with a rather complicated structure (high temperature and pressure in the core). Where would this material

Fig. 22-5 Nebula M8 in the constellation Sagittarius. Notice the small dark spherical dust clouds, thought to be stars early in stage *0-2* of Figure 22-1. [Lick Observatory photograph.]

come from? It has to be gas not already in a star. Is there any such material available?

Not all of the material that we can see in the universe is in stars (spheres of gas); some of it is in nebulae (thinner clouds of gas). Figures 21-9 and 22-3 show nebulae associated with individual stars and probably ejected from them. But there are also very much larger clouds of gas, like the one shown in Figure 22-4, glowing as they absorb and reemit the light of stars which happen to be nearby. Their dark edges and irregular dark patches show that there are some nebulae not illuminated by nearby stars. These dark nebulae look like clouds of smoke. As we will see later, they are clouds of dustlike particles.

In 1946, Bart J. Bok, then at Harvard College Observatory, found many such dark clouds of dust in the nebula shown in Figure 22-5, and even smaller dark spheres of dust here and there in this and other nebulae. These small "globules" of dust are believed to be stars starting to form. In a nebula, atoms and molecules of gas and particles of dust slowly begin to collect together, attracted by their gravitational pull on each other (inset to Fig. 22-1, stage *0-1*). As one of these globules gets larger, it exerts more gravitational attraction, and more and more material is attracted to it. The pull of gravity toward the center gives the collection a spherical shape, perhaps half of a light-year in diameter. In time, gas and dust from a large region of the nebula falls into the globule and becomes part of it.

Slowly the globule contracts because of its own gravity. This contraction increases the pressure and temperature inside, where it becomes so hot that the dust also becomes gaseous. It gets so hot near the center that the gas radiates, and energy begins to flow outward, being absorbed and reemitted, until it reaches the surface, now about 10 or 20 AU away. A star is born as a glowing globule, and stage *0-2* (Fig. 22-1) begins.

The mass of the new star depends on the amount of material it collected during stage *0-1*. This is determined not only by

Fig. 22-6 A cluster of small bright stars associated with nebulae, photographed by G. H. Herbig and G. Haro. [Lick Observatory photograph.]

the amount of dust and gas in the region of the nebula where the star began to form, but also by how fast these particles were moving. Some may have had speeds above the globule's escape velocity, which, of course, is smaller for particles farther from the globule center. All the characteristics of a star, as we have seen, depend on its mass. Mass determines luminosity, spectral type, and surface temperature during a star's long stay on the main sequence, as well as the length of its life before it becomes a white dwarf.

In time, the contraction of the star raises its central temperature and pressure high enough so that the formation of helium from hydrogen can begin. Then it is a main-sequence star, about to begin its development through stages *1* to *6* (Fig. 22-1). The contraction stage is a relatively short one (Table 19 and Fig. 22-2). In addition, the young stars have very low luminositites for much of the stage (Fig. 22-1). Therefore, not many of them would be visible, even through large telescopes. Developments in stage *0-1* would take place so quickly, however, that photographs of a region like that in Figure 22-5, made in the next 10 or 15 years, might reveal small changes in the globules.

Figure 22-6 shows a small cluster of very young stars, bright enough to light up the wisps of nebula which remain near them. They are probably in stage *0-2*. Photographs like this are one kind of evidence that the stars in a cluster all formed at the same time. Since the stars in any one cluster are all the same age (yet can be of different masses), a study of the stars in clusters should tell whether the theory of stellar evolution is correct.

NGC 2264 (Fig. 22-7) is a small open cluster in a cloud of gas and dust. Only its *A* stars have reached the main-sequence stage. Other, less massive ones are still in stage *0-2*.

Now look at Figure 22-8, an H-R diagram typical of the globular clusters. The main sequence has "burned down" farther than in any open cluster—almost to the

Fig. 22-7 The very young cluster NGC 2264 near the "Cone nebula." [Mount Wilson and Palomar Observatories.]

G stars. The globulars must be the oldest type of cluster. There are many red giants and, in addition, stage *3* is represented. In one globular cluster a planetary nebula (stage *4*) has been found. No white dwarfs have been found in globular clusters, but all these clusters are so distant that a white dwarf could not be seen.

The stars in clusters are, in fact, just what the present theory of stellar evolution predicts. And it must be confessed that astronomers had one eye on the cluster stars as they worked out their theory.

In none of the hundred globular clusters known is the main sequence "burned down" below the $G2$ stars. It looks as though no star less massive than the sun has yet had time to evolve away from the main sequence. This means that the universe is not infinitely old. On the contrary, it appears that the universe is less than 20 billion years old—the length of time that it takes a $G2$ star to evolve to the end of the main-sequence stage. All the stars beyond this stage have developed from more massive stars.

Figure 22-1 is a prediction of our sun's future. The sun will leave the main-sequence stage and evolve to a red giant in some 15 billion years. Then its photosphere will reach surely beyond the orbit of Mercury and perhaps even to the earth. Its luminosity will increase a hundred times. The oceans of the earth will

175

Lives of the Stars

boil away, and the baked surface will become as lifeless as the moon's. Eventually the sun will leave the red-giant stage. Perhaps it will go through a stage of variability and explode as a nova or supernova. Finally it will become a white dwarf, as small as one of its planets. By then, the main-sequence *K*- and *M*-type stars, which are today less luminous than the sun, may still be shining, but a new generation of *O, B, A,* and *F* stars will be on the upper part of the main sequence.

Life will be gone from the earth 15 or 20 billion years from now, but will it be gone from the universe? Sometime in the past history of our sun, probably before

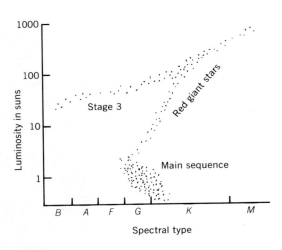

Fig. 22-8 H-R diagram typical of a globular star cluster.

it reached the main-sequence stage, it acquired a family of planets. The stars show such similarity in their life stories that it would be odd indeed if our sun were the only star to have planets. In fact, if the nebular material contracting to form a star were rotating slightly, it would flatten into a spinning disk with a denser "hub" near the center. Chunks of the disk could become the planets, revolving around the "hub" (sun) in almost the same plane, and in the same direction, just as in our solar system.

Until 1950, many astronomers thought our solar system was a freak, caused by another star sideswiping the sun, but a recent theory shows that planets would form in a spinning disk. Since it is likely that many globules would have some slight rotation (and therefore contract into spinning disks), there probably are planets revolving around many stars. In some of these systems there could be a planet at the proper distance from its "sun," and of the proper size to have oceans and an atmosphere like ours on the earth. If so, life could develop on these planets. There may be millions of them, so the universe may always have living things on the planets of other stars, even after life is a thing of the past on the earth.

We could not expect to see any of these planets. They would have to be not much larger than Jupiter—otherwise they would contract into spheres of hot gas, forming double stars (that we do see). Even at the distance of the nearest stars, a non-luminous body of Jupiter's size could not be seen. However, in 1963, Peter van de Kamp of Swarthmore College discovered from its motion that Barnard's star (the second nearest star to the earth) has an invisible companion. He determined its mass and found it to be 0.0045 that of the sun—less than twice that of Jupiter. So the companion is not massive enough to be luminous, and it may well be the first planet of another star to be discovered.

Additional Reading

JASTROW, ROBERT, *Red Giants and White Dwarfs, the Evolution of Stars, Planets and Life:* New York, Harper and Row, 1967.

PAGE, THORNTON, ed., *Stars and Galaxies:* Englewood Cliffs, N.J., Prentice-Hall, 1962, pp. 43-78.

PAYNE-GAPOSCHKIN, CECILIA, *Stars in the Making:* Cambridge, Mass., Harvard University Press, 1952.

SANDAGE, ALLAN, "The Birth and Death of a Star" in *Astronomy* (Samuel Rapport and Helen Wright, eds.): New York, New York University Press, 1964.

chapter 23 | Our Galaxy, The Milky Way

Everyone who has looked at the stars on a clear, moonless night has seen the Milky Way, crossing the darker sky like a faintly glowing ribbon. This band of milky light forms a ring around the celestial sphere, passing through the constellations Sagittarius, Cygnus, Cassiopeia, Perseus, Auriga, and Orion, then through Puppis and Centaurus (not visible as far north as the United States), and back to Sagittarius (Fig. 2-6 a, b, c, d).

The Milky Way looks different from the rest of the sky, and Galileo found out why. His telescope showed it to be made up of stars (Fig. 7-4) that are much fainter and much closer together than those in other parts of the sky, so that they blur together when you look at them. But why should there be so many stars and such faint ones in this particular band around the celestial sphere?

In 1750 Thomas Wright, an English telescope maker, saw that if all the stars were in a thin disk, like a grindstone, with our sun down in the center of the disk, the Milky Way would be explained. Standing at the center of this disklike

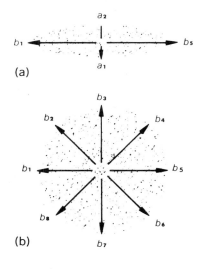

Fig. 23-1

universe, we would see fewest stars along directions a_1 or a_2 (Fig. 23-1a) if the stars are distributed evenly through the disk. Along directions b_1 or b_5, however, the disk extends for a greater distance. If our line of sight is along b_1 or b_5, therefore, we would see many more stars.

Some 30 years later, Sir William Herschel decided to test Wright's idea by counting the number of stars in many areas, scattered over the sky, and estimating the distance to the farthest star that he could see in each. This he did by assuming that all the distant stars are equally luminous, then measuring their brightnesses (p. 150), and using the formula $R = \sqrt{L/b}$. By 1785 he had studied almost 700 different regions. In some of them, the field of view of his telescope showed only a single star, in others as many as 600, and the distribution agreed with Wright's prediction. Herschel's star counts and distance measures showed that the sun is inside a great disk-shaped system of stars, which came to be called the *galaxy* (from the Greek word for milk). He also saw that the large, glowing nebulae, like those in Figures 22-4, 22-5, and 22-7, lie in the Milky Way and help to produce its soft and beautiful light.

Herschel thought that there was a long deep cut (or rift) in the margin of the galaxy. Figure 23-2 shows the appearance of the sky in the direction of the deep indentation. The rift runs lengthwise along the Milky Way for some distance beyond the edges of the photograph and cuts the Milky Way in two. It does look as though the galaxy ends just beyond the foreground stars in Figure 23-2; and up to less than a century ago, astronomers thought that these rifts were like cracks in

Fig. 23-2 A portion of the Milky Way in the constellation Cygnus, showing the dark "rift." [Mount Wilson and Palomar Observatories.]

the "grindstone" through which we could see out into the dark empty space outside the galaxy.

But these dark areas cannot be empty space because they are not absolutely black. Even the darkest of them produces a little light — enough to be measured with modern equipment. They turn out to be clouds of dust particles that cut off our view of the stars beyond like a dark curtain. They cannot be clouds of cool gas because gas is transparent unless it is extremely dense. So great a mass of gas would be needed to cut off the starlight beyond that nearby stars would move in orbits around it. But a cloud of smoke (very small dust particles) is very efficient in absorbing light. A relatively small mass of smoke can easily cut off the light of the stars beyond. Each particle absorbs some of the light falling on it (and reemits the energy as invisible heat waves). The particles

Fig. 23-3 The "Horsehead" nebula in Orion, a photograph of 60-minutes exposure time made by Georges Van Biesbroeck with the 82-inch reflecting telescope of the McDonald Observatory in Texas. The head of the "horse" is a dark dust cloud with bright edges. [Yerkes Observatory photograph.]

also reflect part of the light, sending it out in all directions. Starlight, which had been coming straight toward us, is now scattered in many directions and we no longer see it as parallel beams of light from a star. Some of the reflected light bounces off other particles, giving a faint glow to the dark cloud, especially near stars close to it.

The dark area in the constellation Cygnus (Fig. 23-2) is one of these dust clouds, or dark nebulae. It cuts off the light of all the stars beyond it. Because it is nearer than the other dark nebulae, it cuts off the light of more stars than they do, and caused Herschel to think that the galaxy does not extend very far outward there. Another dark nebula is shown in Figure 23-3. The small dark clouds silhouetted against the bright nebula in Figure 22-5 (near the new-born stars) are also dark nebulae. Like the bright gaseous nebulae (Figs. 22-4, 22-7), these clouds of "smoke" are found in the Milky Way. So it is not surprising that most of the highly luminous young stars of the main sequence are found here too. These dark clouds are the birthplace of the stars.

Fifty years ago, astronomers were puzzled by the spectra of distant stars, whose spectral lines indicate that they are hot stars of spectral type B — blue stars which have the peak of their Planck curves at about 3000 A. These distant stars, however, appear red, like cool stars of spectral type G, with maximum radiation at 5000 A. Laboratory experiments explained this by showing that when light shines through a cloud of widely separated, very tiny dust particles (about 0.0005 of an inch in size, like smoke particles), more of the violet and blue light is absorbed than the red and infrared. A greater percentage of the light of longer wavelengths comes through; the amount of absorption is inversely proportional to wavelength. (Long radio waves come through unaffected—a fact of great importance.) The light looks redder and the light source looks dimmer through smoke, as you can check for yourself by looking at the moon or distant lights through smoke or smog.

The reddening of the light of distant stars means that there is dust, not only concentrated in dark nebulae, but spread more thinly between the stars throughout the galaxy. The farther a star's light travels, the more dust it meets and the more it is dimmed. The effect of the dust is not apparent in the light of fairly nearby stars (up to 100 parsecs). The amount that a distant star's light is dimmed can be determined by comparing the Planck curve expected from the star's spectral type with the radiation measured in the star's spectrum—the star's color. At each wavelength in the Planck curve, some fraction of the light is absorbed and this fraction increases for shorter and shorter wavelengths ($\propto 1/\lambda$). If we assume that the interstellar dust is spread uniformly through space, the distances of far-off stars can also be estimated from the amount of this dimming. Because more blue light is absorbed than red light, stars seen through dust are always reddened. How fortunate it is that Miss Annie J. Cannon had classified the main-sequence stars by the absorption-line pattern in their spectra, rather than by their colors!

In laboratory experiments, the amount of dust per cubic mile needed to produce the observed dimming in stars was found. Throughout the galaxy, between the stars and nebulae, there are about 100 tiny dust particles in each cubic mile—interstellar space is not really crowded. The dark nebulae, that cut out the light almost completely, may contain only a hundred times as many particles in the same volume. Nevertheless, were it not for the dimming effect of this dust in space, we would be able to read at night by the light of billions of stars in the Milky Way.

In addition to the dust (probably specks of carbon and frozen hydrogen, oxygen, and carbon compounds, such as ice, carbon dioxide, and methane), there is also gas (chiefly hydrogen and helium) in the interstellar material. The presence of some gases is shown by faint absorption lines which they put into the spectra of distant stars. Hydrogen is detected by its radio emission. The amount of gas is estimated at one atom per cubic centimeter (earth's atmosphere at sea level contains 10^{19} molecules in the same volume) and perhaps as many as 100 atoms per cubic centimeter in the bright nebulae. Estimates of the amount of gas and of the stars' masses show that the material of the galaxy is about equally divided between stars, on the one hand, and nebulae and interstellar material, on the other.

During the nineteenth century, astronomers realized that the galaxy would not end as abruptly at its border as Figure 23-1 indicates. The stars, almost uniformly distributed over most of the galaxy, would thin out gradually near the edges. They tried to place the edge of the galaxy by finding where the stars began to thin out. Better and better telescopes allowed them to see farther and farther. However, the presence of interstellar dust was still undetected because almost nothing was known about spectral types. Dimming by this dust makes distant stars seem much fainter, and therefore, farther away than they are. Since the effect increases with increasing distance, this made the galaxy's stars appear to thin out much nearer to us than they actually do. Until early in this century, the accepted estimate of the galaxy's diameter (b_1 to b_5, Fig. 23-1b) was 10,000 light-years. When the dimming by interstellar dust was taken into account, the edge of the galaxy (where the stars thin out noticeably) was estimated to be 20,000 light-years (4500 parsecs) from the sun.

Estimates in the direction of Sagittarius (Fig. 7-4) were more difficult to make, since the stars are much more thickly crowded there. However, since the sun was assumed to be at the center of the galaxy, measurements made in the opposite direction, toward Gemini, Auriga, and Perseus, would give the radius (20,000 light-years). The diameter of the galaxy would then be 40,000 light-years.

The fact that the distant stars of the Milky Way are most dense near Sagittarius might suggest that the galaxy extends farther in that direction, and that the sun is not at its center. Perhaps it did suggest this to some astronomers. But it must have appeared unlikely, just as the idea that the earth might not be at the center of the universe had appeared unlikely some four hundred years earlier.

In 1917, Harlow Shapley, by then at Harvard College Observatory, became interested in the region of the sky near Sagittarius, where he could count 93 globular clusters. In a few of the nearer clusters he could see variable stars that he thought were cepheids. He determined the luminosities and distances of these variable stars from their periods, using the method he had devised a few years earlier. This gave him the distances to these few globular clusters. He observed the total brightness of each of these clusters and then calculated its total luminosity ($L = b R^2$). It turned out to be about the same for all of them: 200,000 times the sun's luminosity. It seemed reasonable that the luminosity of the other, more distant ones, would also be the same. He measured the brightness of each of the other clusters and, since $R = \sqrt{L/b}$, calculated the distances of even the faintest ones. He also carefully measured their positions on the celestial sphere.

He found that they were clustered in a "bubble" about 20,000 light-years in radius, and extending well above and below the "grindstone" or galactic disk. He calculated that the center of this sphere, or bubble, lay 130,000 light-years from the sun in the direction of Sagittarius. Yet the opposite edge of the galaxy (in Auriga) had been measured as only 10,000 light-years from the sun (later corrected to 20,000). This meant that our sun is not at the center of the galaxy! Shapley's original estimate of the distance to the center of the sphere of clusters has now been revised to 30,000 light-years, the estimate used in Figure 23-4a. It still places our sun off center.

Shapley believed that the center of the sphere into which these clusters were packed was also the center of the galaxy (its nucleus, as shown in Fig. 23-4b and 23-4c). This made the galaxy's diameter about 300,000 light-years (now revised to 100,000 light-years, or 30,000 parsecs). This was a startling idea. Was there other evidence to back it up?

Newton's laws tell us that if the planets were not moving, they would fall into the sun. Because they are moving, they revolve around the sun, always falling toward it but never falling into it (p. 58). If the stars and nebulae and interstellar

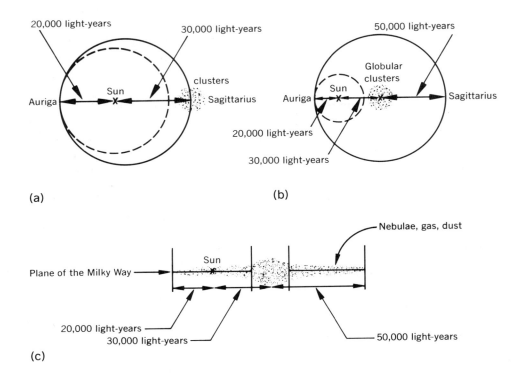

Fig. 23-4 (a) Shapley found the sphere of globular clusters centered at 30,000 light-years (revised estimate) from the sun in the direction of Sagittarius. Since the edge of the galaxy in Auriga is 20,000 light-years (revised estimate) from the sun, the sun is not at the center of the galaxy. Dashed circle shows previous idea of the galaxy. (b) Shapley believed that the center of the galaxy lies at the center of the sphere of globular clusters (called the galaxy's nucleus), 30,000 light-years (revised estimate) from the sun. The edge of the galaxy is 20,000 light-years (revised estimate) from the sun in the opposite direction. The dashed circle shows the earlier idea of the galaxy. (c) Cross section of the galaxy, according to Shapley. Distances are the revised estimates.

matter of the galaxy were motionless, they would all fall together into one big mass of gas at the center. But the stars are billions of years old (Table 19) and have not fallen together. Therefore, they must be revolving, like a gigantic solar system. Men had always thought of them as the "fixed stars," but when Herschel showed that these stars are gathered together in a disk, it became clear that they must all be moving.

Maps of the celestial sphere drawn by Ptolemy show the stars in the same positions in the sky as you see them tonight. The angular distances between the stars and their directions from each other do not appear to have changed. This might mean that the galaxy is rotating like a big wheel, as the earth does each day, rather than being made up of separate bodies revolving around the center of the galaxy, like the planets revolve around the sun, following Kepler's law, $P^2 \propto R^3$ (slower angular speed for larger R). If the "big-wheel" rotation were correct, every object in the galaxy, including the sun, would move at the same angular speed.

The changes of position in the sky that we are looking for, called "proper motion," are like parallax. Up to now we have considered only yearly parallax, the shifting back and forth of the position of a star on the celestial sphere, caused by the orbital motion of the earth. The kind of parallax that we are looking for now is a steady change of a star's position in the sky, caused by the motion of the star past the sun (Fig. 5-3b). Of course, it could be due to equal but opposite motion of the sun past the star. However, we would expect the proper motions of the stars to be due to the orbits of stars around the center of the galaxy, relative to the motion of the sun (and earth) in its orbit around this center (Fig. 23-5).

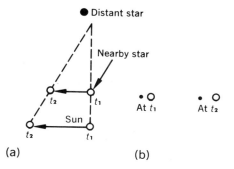

Fig. 23-5 (a) At time t_1 the sun and a nearby star are at positions t_1. By time t_2 they are at positions t_2. (b) The relative positions of the nearby star and the more distant one, as photographed from the earth, change between times t_1 and t_2.

The farther a star is from us, the less will be its angular shift on the sky (per year or decade) for the same relative movement in miles per second (Fig. 5-3c, d). The distant star in Figure 23-5 is so far away that it does not appear to have any proper motion at all, so it is used as a reference point to detect the movement of the nearer star. Therefore, it should not be surprising that proper motions can be detected only for stars fairly close to the sun. In fact, there are only 330 stars whose proper motion is as much as 1″ per year. The star with the largest proper motion is Barnard's star (Fig. 23-6), 1.8 parsecs from the sun, which steadily changes its position in the sky by 10.25″ each year. To a smaller extent, the positions of the stars in the Big Dipper are slowly changing.

These measured proper motions indicate that the stars in our galaxy are moving and suggest that their motions are in orbits, but they do not give enough information to locate the center about which the stars are moving. Astronomers had to look for that in another way.

The motions of the planets are detected by their changing positions in the sky. Because they are so near, the planets' changes of position are much larger and easier to see than those of the stars that we have been considering. However, both motions can be measured in another way—by Doppler shift. Distance does not make the Doppler shift less; it can be read from the absorption lines of any star that is bright enough for its spectrum to be photographed. The Doppler shift in the spectrum of a planet moving toward us at 30 miles per second is the same as that in the spectrum of a star 10,000 light-years away and moving toward us at the same speed. Of course, as we have seen, Doppler shift measures only that part of the motion which is toward or away from the observer. However, as Mercury circles the sun, or one star in a pair moves around the pair's center of mass,

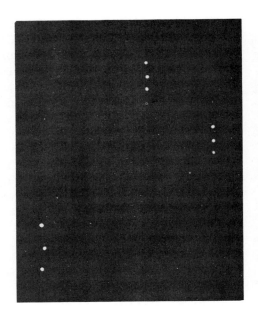

Fig. 23-6 Photographs of Barnard's star (lower left) and of two more distant stars (upper right) were taken on September 1, 1948, March 20, 1949, and September 8, 1949. The three photographs were then printed on one plate, offset so that there are three images of each star. Notice that the three images of Barnard's star are farther apart than the three images of the other stars. This shows that Barnard's star has moved with respect to them. Notice that in the middle image, Barnard's star has also moved slightly to the right. This is due to another of its motions — in an orbit around its unseen, jupiterlike companion (p. 176). The proper motion of Barnard's star (10.25″ per year) is larger than its yearly parallax (0.54″). This is true of many stars which are near enough for parallax determinations of their distances to be made, and adds to the difficulty of this measurement. [Courtesy Peter van de Kamp, Sproul Observatory.]

there will be a Doppler shift whenever the motion is not exactly across our line of sight. This small shift in spectral lines measures the part of the star's motion that is toward or away from us, called the *radial velocity*.

In the 1920's, Bertil Lindblad of Sweden and Jan H. Oort, a Dutch astronomer, studied the radial velocities of thousands of stars in all parts of the Milky Way. Toward Sagittarius the radial velocities average zero, as they do 180° away from Sagittarius (near Auriga), 90° away (in Cygnus), and 270° away (near Puppis). However, they found large average velocities of approach for stars 45° either side of Cygnus (in Aquila and Monoceros) and large velocities away from us for stars 45° either side of Puppis (in Perseus and Centaurus). All these different velocities can be explained if the center of the galaxy is toward Sagittarius (at *C* in Fig. 23-7), and if all stars are orbiting around the center like the planets orbit around the sun. Right now, the sun (at *S* in Fig. 23-7) is moving at a speed of about 200 miles per

Fig. 23-7 Averaged radial velocities of the stars show that the center *C* of the galaxy coincides with the center of the sphere of globular clusters. Arrows show average orbital motions of stars around *C*. Directions from the sun *S* are indicated by names of constellations in the Milky Way.

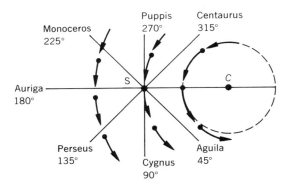

second toward the contellation Cygnus. This can be determined by a consideration of the orbits (Fig. 23-7) and a careful mathematical study of the measured radial velocities from the Doppler shifts of many stars at known distances and known directions from us.

The mass of the galaxy can be determined from Kepler's law in the same way as the masses of double stars, with the difference that because the mass of the sun is so tiny a fraction of the whole galaxy, we can leave out its mass (just as we could omit a planet's mass when computing the mass of the sun). Since R is 3×10^4 light-years, and there are 6.3×0^4 AU to a light-year, $r = 2 \times 10^9$ AU. The period is 2×10^8 years. Therefore, $m_{galaxy} = (2 \times 10^9)^3/(2 \times 10^8)^2 = 8 \times 10^{27}/4 \times 10^{16} = 2 \times 10^{11}$ (200 billion) times the mass of the sun. Since the sun is an average star and about half of the material of the galaxy is in the form of stars, we can conclude that the galaxy contains about a hundred billion stars!

Until about 1925 it seemed likely that the proportions of the different types of stars is the same everywhere in the disk of the galaxy as they are in the small region near us (Table 19, col. 4). Then Walter Baade of Mount Wilson Observatory showed that the O and B stars (young, luminous, and massive) are confined to the central plane of the galaxy—confined to the Milky Way in our sky (Fig. 23-4c). Even in this plane, they are not uniformly distributed. Painstaking measures of their distances showed that these young stars and the nebulae (bright and dark clouds) are in three bands or "arms"—parts of circular rings in the plane of the Milky Way. The other types of stars are not confined to the arms, although they are most thickly distributed there.

Within these bright arms, stars are being formed; the material is there, and the young O and B stars are evidence that this material has been used within the last few million years (recently, as time goes in the galaxy). Are the arms permanent features? Older stars—less massive main-sequence ones, red giants, variable stars, novae, planetary nebulae, and white dwarfs—are found all through the disk of the galaxy. This suggests that arms (in which, presumably, they formed) have not always been in the same places among the stars. How do arms form? How do they disappear? No one knows—yet.

From deep within the galaxy, astronomers have learned its shape, its size, its mass, its age, its structure, and the number of stars it contains. The sky they have studied is the interior of the galaxy—a majestic sight on a clear, dark, moonless night. But think how magnificent our Milky Way galaxy must look from the outside, in the gleam of billions of stars and glowing nebulae.

Additional Reading

Bok, B. J., and P. F. Bok, *The Milky Way:* Cambridge, Mass., Harvard University Press, 1957.

Millman, P. M., "The Dust and Gas of Space" in *Astronomy* (Samuel Rapport and Helen Wright, eds.): New York, New York University Press, 1964.

| **Beyond the Milky Way**

One of Wright's ideas touched off an argument that was not settled until 1924. He believed that our galaxy is not the only island in the sea of space; he said that there are many others and that he could point to them in the sky. This possibility was more startling and difficult to accept than his other suggestion that the stars are gathered into a disk-shaped system. It was also much more difficult to prove or disprove.

Here and there among the stars, small faint blurs of light can be seen. Each one looks larger than a star, like a dim, roundish cloud. Beginning with Galileo, telescopes had shown that some of them are clusters of stars. As more observations were made with larger telescopes, more and more of these clouds were discovered and many proved to be star clusters. The majority, however, still appeared as roundish or oval clouds in which no individual stars could be seen. Wright believed that these were other universes of stars, like our galaxy, but far beyond its borders. We see them between its stars, he said, much as the outline of a distant forest can be glimpsed between nearby trees. Astronomers were not convinced that they are really "distant forests"; it seemed more likely that they were merely "thickets of nearby bushes," as the globular clusters had proved to be.

A German philosopher, Immanuel Kant, independently arrived at Wright's conclusion five years after he did. Kant was famous and respected, and his presentation was so clear and sensible that astronomers considered the possibility seriously and tried to find evidence that would settle the matter.

In 1781 a French comet hunter named Charles Messier prepared a list of 57 star clusters and 45 "clouds" (together with their "latitudes" and "longitudes" on the celestial sphere), so that they would not be mistaken for comets. They are known today by their numbers in his catalogue. The Hercules star cluster, for instance, is called M13 (it is the thirteenth object on his list). Other astronomers found many more "clouds" and a few more star clusters. With characteristic energy, Sir William Herschel, aided by his son John, added 4630 more. In 1864, Sir John published the complete list as the *New General Catalogue*. The objects he listed there, and those added later, are known today by their numbers in this catalogue. For instance, M13 is also called NGC 6205; the cluster shown in Figure 22-7 undetected by Messier's telescope, is NGC 2264.

All these observations failed to determine the nature of the "clouds." Sir John Herschel included both star clusters and "clouds" in his catalogue because it seemed likely that, as stronger telescopes were built, all of the "clouds" would prove to be clusters. Sketches made in the first half of the nineteenth century indicated that telescopic observations of the "clouds" showed no more detail than does Figure 24-1. In addition, there was no way to measure their distances. Their apparent lack of parallax merely meant that they are considerably farther away than 65 light-years (20 parsecs).

However, another line of evidence soon showed that many of the unexplained "clouds" are not star clusters. Five years before the catalogue came out, Gustav Kirchhoff's laboratory experiments showed that a cloud of glowing gas produces a spectrum of bright emission lines (Fig. 19-1) and no continuous spectrum. The year that the catalogue was published, spectra of the "clouds" began to be photographed. Unlike the star clusters, whose spectra resemble those of the sun and stars, the spectra of many (like the one shown in Fig. 24-1) contain only bright lines. These particular "clouds" really are clouds. They are nebulae like the larger ones in the Milky Way, whose structure had already revealed their nature. Later, photographs with larger telescopes revealed similar structure in these smaller clouds (M1, NGC 1952 shown in Fig. 22-3). But these early spectra showed that they are made of low-density gas and that they lie within our galaxy. It takes a star to make a nebula glow, by supplying it with the radiant energy to absorb and reemit as light with a bright-line spectrum. In the sky near these neb-

Fig. 24-1 The larger picture of the constellation Orion was taken by Hans Pfleumer of North Brunswick, New Jersey, with his 6-inch Tessar telescope, exposure time 10 minutes. The inset shows the bright blur in Orion's sword (M42, NGC 1976) photographed through the same telescope with a 54-minute exposure time. [Courtesy Hans Pfleumer and *Sky and Telescope* magazine.]

ulae are stars whose brightness makes it clear that they are part of our galaxy. Thus, another larger group of the "clouds" was removed from the list of possible distant galaxies.

Nevertheless, even with the nebulae weeded out, a majority of the "clouds" were left unexplained. Early in this century, it became clear that these had several things in common. Their light shows a continuous spectrum crossed by dark absorption lines, indicating that they are systems of stars. None of them is found in a strip along the Milky Way from 10° to 30° wide. (This is just where we would be least likely to glimpse anything outside the Galaxy, for here the stars are most closely packed in the sky and dust clouds and nebulae form a "curtain.") The radial velocities (p. 184) of some of these "clouds" were measured. They are different from those of stars or nebulae near them in the sky, for no matter where they lie, Doppler shifts indicate that most of them are moving away from us at high speeds.

By this time, large instruments like the 40-inch lens telescope at Yerkes Observatory in Wisconsin and the 100-inch mirror telescope at Mount Wilson Observatory in California, were in operation. Long time-exposure photographs made with them not only revealed the beauty of these objects, but clearly showed details

Fig. 24-2 NGC 4594, spiral galaxy, in the constellation Virgo. [Mount Wilson and Palomar Observatories.]

of structure only glimpsed before. M31 shown on the cover, like many of the others, looks like a glowing pinwheel of light; it has spiral "arms." Because of this, the "clouds" now began to be called "spirals."

Fresh from your study of Chapter 23, you probably have already noticed the strong resemblance between M31 and our galaxy. M31 looks flat and disklike. There are spiral arms and there is a nucleus at the center, resembling the nucleus which Shapley found in the Milky Way galaxy. Unlike our round "grindstone," M31 is oval in shape. It could be that it is tilted and that others look round because they lie at right angles to our line of sight. We see still others, like the one in Figure 24-2, in profile view.

But back in 1917, Shapley was just announcing his idea of the galaxy nucleus of tightly packed stars surrounded by globular clusters. It was not until 20 years later that there was enough information about our galaxy to draw its profile. It was 25 years later that even a trace of spiral arms was found in our galaxy and it was 40 years before the 21-centimeter radio studies confirmed the presence of these arms. The geography of the Milky Way galaxy was pieced together from inside observations at the same time as the nature of the spirals was being investigated by observations of their outside appearance. Both studies aided each other by suggesting what to look for, but there were no definite conclusions from one study that could be used to prove the other.

At that time, most astronomers agreed that spirals are disk-shaped systems of stars. Opinion, however, was divided as to whether they are fairly small systems within our galaxy, or much larger ones outside it. The only way to settle the question was to find out how far away they are.

In some of the spirals a few faint individual stars could be observed. Series of photographs made at the Mount Wilson Observatory showed that several of these stars had brightened suddenly, like novae. The brightness of each of these faint novae was measured. If their luminosities were about the same as those of novae that had occurred in our galaxy, then (according to $R = \sqrt{L/b}$), they are about a million light-years away — well beyond the galaxy's limits — and so are the spirals that contain them. It was quickly pointed out, however, that two much brighter novae (like the later one shown in Figure 24-3) had previously been observed in spirals. In 1888 one in the Andromeda spiral, M31, had temporarily become so bright that it could be seen with a good pair of binoculars. This larger brightness reduced the distance to well within the boundaries of the galaxy, if the luminosity was assumed to be the same as that of ordinary novae in our galaxy. (Much later, after the argument about the spirals was settled, these were found to be supernovae which have a very much higher luminosity than novae.) To add to the confusion, measurements showed that if the luminosity of the brightest individual stars that could be seen in certain spirals is the same as that of the most luminous stars in our galaxy, they are over a million light-years away and lie in distant galaxies.

And so the argument went. It climaxed in a sort of astronomical Lincoln-Douglas debate between Harlow Shapley and H. D. Curtis, an astronomer from Lick Observatory, held in 1920 at the National Academy of Sciences in Washington. Shapley believed that the evidence showed the spirals to be within our galaxy; Curtis defended his belief that they are galaxies like ours.

The debate did not settle the question, of course. It had to be settled at the telescope — and four years later it was. Then, Edwin Hubble of Mount Wilson Observatory discovered variable stars in three of the nearer spirals — among them M31 (cover). Their light curves (like that in Fig. 21-4) showed him that they

 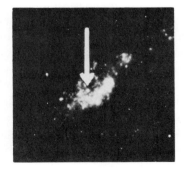

Fig. 24-3 Supernova in NGC 5457. The photograph to the left was taken on June 9, 1950 (before the star flared up); that to the right was taken on February 7, 1951. Arrow points to the supernova. [Mount Wilson and Palomar Observatories.]

are cepheids like those in the Magellanic Clouds that Shapley had studied 10 years before (p. 162). Comparing their luminosities (revealed by their periods) and their brightnesses ($1/10^8$ that of the sun), he found that these cepheids are indeed remote—far beyond the confines of our galaxy. Those in M31 are 10^6 (a million) light-years from the sun—and therefore, so is M31. Thirty more spirals with recognizable cepheids were found; the cepheids showed that the distances of these spirals are between 1 and 20 million light-years.

Now the long search, on which Wright and Kant had set astronomers 175 years before, was ended and a whole new field of astronomy was opened. The frontiers of space had been extended and it was soon apparent that the universe is much larger and more complicated than anyone had supposed.

Some galaxies with cepheids also contain O- and B-type stars and globular clusters, and ordinary novae have been observed in some. If the luminosities of these objects are assumed to be about the same as the O and B stars, globular clusters, and novae in our Galaxy, their distances come out about the same as those calculated for the same galaxies from cepheid periods. This showed that these other objects can be used to establish the distances of galaxies in which cepheids cannot be seen. In this way, galaxy distances up to 80 million light-years were measured.

If the distance of a galaxy is known, its size can be determined from its angular diameter in the sky (Fig. 14-2). The diameter of M31 (cover), 10^6 light-years away and with an angular diameter of about 2°, is 40,000 light-years, not too different from that of our galaxy (50,000 light-years). M31 and NGC 5457 (Fig. 24-3 somehow look as if they are "rotating"; studies of the Milky Way galaxy suggest that the nebulae and stars in each are moving in orbits (p. 181) around the center of its galaxy. Inclined lines in the spectra of about 25 galaxies indicate that they are. From the radial velocities (recorded as Doppler shifts in spectra like this) astronomers can calculate the orbital speed v of a star or nebula around the galaxy's center. The distance (R, in AU) of any of these stars or nebulae from its galaxy's center can be determined. Since the circumference of its orbit is $2\pi R$, the period ($2\pi R / v$) can be determined. In M31 it turns out to be about 200 million years for stars or nebulae well out in the arms.

The mass of a galaxy can be determined from these orbital speeds and distances from the galaxy's center, using Kepler's law, much as the mass of our galaxy was determined. The Andromeda galaxy, M31, is 3×10^{11} times that of our sun—not too different from that of our galaxy (2×10^{11}). Since the proportion of nebulae to stars seems similar to that in the Milky Way galaxy, we can conclude that the Andromeda galaxy contains about 1.5×10^{11} stars. Masses can be determined in this way for only about 25 galaxies. The spectra of the others are too faint and their rotational speeds too slow to make the Doppler shifts measureable. However, many galaxies come in pairs that revolve around a center of mass. These speeds are larger and easier to measure and indicate the mass of the pair in much the same way as do the motions of double stars.

The largest telescope in the world, the 200-inch reflector at Mount Palomar (competed in 1948) was built especially to observe galaxies. It not only can "see" more detail in nearby galaxies, it can also "see" farther than other telescopes. Galaxies extend as far as this giant eye can see. On photographs made with the Palomar telescope pointed away from the Milky Way, there appeared tiny images of galaxies that must be very far away and more numerous than the images of foreground stars of our galaxy. Hubble was able to estimate from a survey of samples of the sky that there are a billion of them within the range of the Palomar telescope.

Of course it is clear that the galaxies that form these tiny images are very far away. But how far? How big is the known universe? If it were certain that the galaxies are all about the same size and have about the same luminosity, then the distances could be obtained by comparing their brightnesses. But Hubble found that they are not all alike, and he classified the different types of galaxies into four main groups: "normal spiral," "barred spiral," "elliptical," and "irregular."

The "normal spirals" (S) seem to grade into each other from type S0 (flattened systems with no trace of spiral structure) through tightly wound Sa galaxies and the looser armed Sb class to the very open Sc galaxies (Fig. 24-4). The arms of Sa galaxies are smooth and featureless; the Sb type shows more structure; and in Sc the arms are broken up into fine detail. Hubble coined the name "barred spirals" (SB) for galaxies like those shown in Figure 24-5. From SB0 through SBa, SBb, and SBc there is a strengthening of the arms relative to the central "bar," and the texture of the arms becomes more broken.

Spirals and barred spirals together account for almost 80% of observed galaxies. They range in diameter from 20,000 to more than 150,000 light-years and in mass from 10^{10} to 3×10^{11} times the mass of the sun. Our galaxy and M31 are thus relatively large as spirals go.

Much fainter than the spirals and barred spirals are the elliptical galaxies. They contain no trace of spiral arms and are not flattened like disks. They resemble the nuclei of spiral galaxies, and their stars are older types, like those in the nuclei of spirals. They account for about 17% of observed galaxies; but their actual proportion may be much higher, as many are only very faintly luminous, making them difficult to see if they are very far away. Examples are the two companion galaxies of M31 (cover), one adjacent to the upper right portion of M31 and one some distance from its upper left margin in the photograph. Ellipticals have a much greater range in size and mass than do spirals. Some are very much smaller than the Milky Way galaxy and some very much larger.

The "irregular" galaxies are formless in appearance, containing lots of dust, and generally smaller and less luminous than the other types. They account for only about 2.7% of observed galaxies. Examples are the Magellanic Clouds (Fig. 21-6). The distance to the Clouds is 150,000 light-years. After Shapley worked out this distance from the periods of the cepheids they contain (p. 163), he overestimated the diameter of the galaxy at 300,000 light-years. Thus, their location (near Auriga, Fig. 23-4b) placed them at the edge of the galaxy. Now they are known to be separate galaxies, rather than star systems within our galaxy.

The fact that there are different types of galaxies, and that so many of them appear to grade into each other in appearance suggests that these types may be states in a galaxy's development, just as the different types of stars are stages in

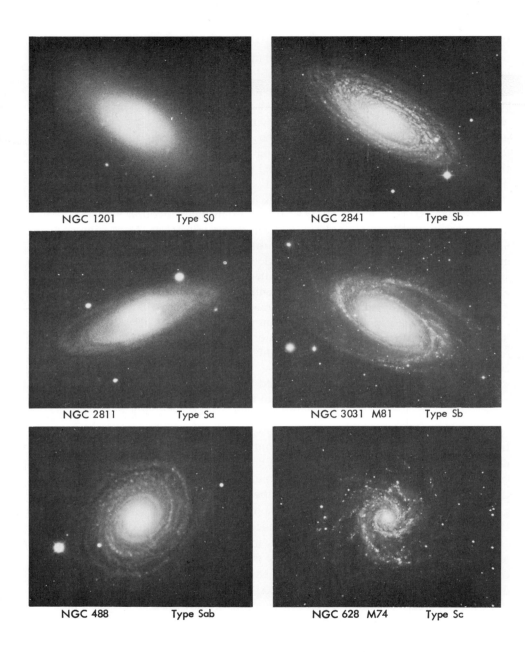

NGC 1201   Type S0   NGC 2841   Type Sb

NGC 2811   Type Sa   NGC 3031 M81   Type Sb

NGC 488   Type Sab   NGC 628 M74   Type Sc

Fig. 24-4 E. P. Hubble's classification of normal-spiral galaxies. [Mount Wilson and Palomar Observatories.]

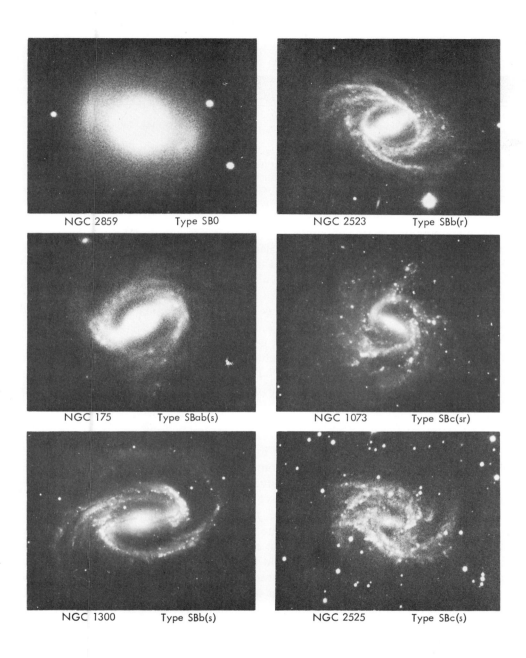

NGC 2859   Type SB0

NGC 2523   Type SBb(r)

NGC 175   Type SBab(s)

NGC 1073   Type SBc(sr)

NGC 1300   Type SBb(s)

NGC 2525   Type SBc(s)

Fig. 24-5 E. P. Hubble's classification of barred-spiral galaxies. [Mount Wilson and Palomar Observatories.]

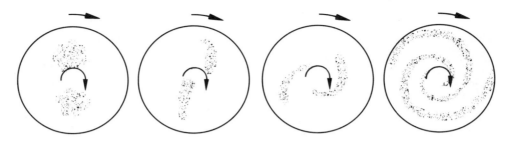

Fig. 24-6 It is not surprising that gas and dust (from which new stars form) is concentrated into spiral arms. Two irregular masses of gas and dust are shown in the drawing at the left. The parts closest to the center of the rotating galaxy move fastest. ($a \propto F \propto 1/r^2$), while those farthest away trail behind.

the development of a star (Chapter 22). Thirty years ago many astronomers felt sure that elliptical galaxies gradually flatten, develop spiral arms, and become spiral galaxies. Later on, they reversed the direction of evolution, assuming that all galaxies begin as irregulars and evolve either through the stages of spirals (Fig. 24-4) or barred spirals (Fig. 24-5), in which the arms become more tightly wound (Fig. 24-6) until finally their gas and dust is completely condensed into stars and they become ellipticals.

Today, most astronomers doubt that galaxies evolve from one type to another. The fact that different galaxies are flattened by different amounts, they say, results from their different rates of "rotation." That is, the faster the rotation, the more likely a mass of gas will become flattened into a disk and the more tightly wound its arms will become (Fig. 24-6).

Nevertheless, the stars within galaxies will evolve. Elliptical galaxies may have always been elliptical, but they may have had *O, B, A,* and *F* stars when they were young. As in a star cluster, they may have burned down like candles (p. 174). Spirals may never become ellipticals, but in time their spiral arms will disappear when all the dust and gas has been converted into stars. Other changes can take place. The S0 galaxies (Fig. 24–4) have the disk shape of spirals but no arms. They are found, in almost every case, in clusters of galaxies. The galaxies in clusters are moving in orbits around the center of the cluster, much as the stars in a star cluster do. It is possible that many of them have collided with other galaxies in the cluster. This may sound like a real catastrophe, but the billions of stars in a galaxy are so widely separated that it is unlikely that any two stars would collide. The much larger nebular interstellar dust clouds, however, would be completely swept out of the galaxies as they pass through each other, and would be left as a third, gassy object somewhere between them. When there is no gas and dust left, no new stars can form. The stars remaining in each of the colliding galaxies would age, and they would become S0 galaxies.

Our galaxy is in a group or small cluster of 17 nearby galaxies, spread over a region 3×10^6 light-years across. The two largest members are both spirals (the Milky Way and M31). There is one more spiral, four irregulars (among them the

Magellanic Clouds), and 10 ellipticals. Of course, there may be other members hidden from our view by the dust clouds and nebulae in the plane of the Milky Way, or fainter galaxies not yet noticed on photographs. Our galaxy is moving with respect to other members of this "local group." From Doppler shifts, we find that the Large Cloud of Magellan is moving away from us at a speed of about 165 miles per second and that M31 is moving toward us at about 155 miles per second, but much of this motion is due to the motion of the sun around the center of our galaxy. After the sun's motion is subtracted, the true motion of the Milky Way galaxy among the others in the cluster remains.

In fact, these motions of galaxies in a group or cluster remain a puzzle. If they are orbital motions around some center of mass, they show that the cluster or group has much larger mass than the sum of the estimated masses of all the galaxies in the cluster. No one yet knows how the galaxies got started in motion, or whether they will remain clustered in groups forever. Perhaps the groups and clusters break up and re-form, like people milling around at a party.

Additional Reading

BURBIDGE, MARGARET, "The Life Story of a Galaxy" in *Stars and Galaxies* (Thornton Page, ed.): Englewood Cliffs, N. J., Prentice-Hall, 1962.

MILLMAN, P. M., "Galaxies" in *Astronomy* (Samuel Rapport and Helen Wright, eds.): New York, New York University Press, 1964.

SHAPLEY, HARLOW, *Galaxies*: Cambridge, Mass., Harvard University Press, 1960.

chapter 25 | # The Universe

Edwin Hubble photographed about 1300 sample regions of the sky with the 100-inch telescope at Mount Wilson Observatory, exposing each photograph for an hour. He counted about 44,000 galaxies on these photographs. They showed about the same number (several hundred galaxies) in each square degree of the sky, except in the band of the Milky Way, where dust clouds hide the galaxies behind them. He corrected the total count for the Milky Way gap and for the parts of the sky that he did not photograph and found that it equalled 3×10^6 galaxies. Comparisons of their brightnesses with those of galaxies whose distances had been measured indicated that the most distant ones he could photograph are about 600 million (6×10^8) light-years away. So the estimates from the Mount Wilson photographs are estimates of the number of galaxies within a sphere of this radius.

Then Hubble surveyed the same sample regions of the sky with the 200-inch telescope at Palomar Observatory, again using an exposure time of one hour. Again, he found an equal distribution of galaxies on every side of us. He also saw faint galaxies that did not show up at all on the Mount Wilson photographs. The larger Palomar telescope can photograph objects which are four times dimmer than those which the 100-inch telescope can record. This means that photographs made with the Palomar telescope show galaxies twice as far away — all of them lie within a sphere 1.2×10^9 light-years in radius. The Palomar photographs showed that there are about 2.4×10^7 galaxies in this larger sphere, eight times as many as the Mount Wilson survey revealed. Since doubling the radius of a sphere increases its volume eight times, this meant that up to this huge distance from the earth the galaxies are distributed evenly.

Hubble found that there are 2.4×10^7 galaxies in a sphere with a volume of $(4/3)(\pi)(1.2 \times 10^9)^3$, or 7×10^{27} cubic light-years. The average diameter of a galaxy disk is only about 3×10^4 light-years. This means that the average distance between galaxies is about 10 million light-years. Since the average mass of a galaxy is about 4×10^{44} grams, there are about 1.5×10^{-28} grams of matter in each cubic centimeter of the universe — about as much as a thimbleful of air spread over a volume equal to that of the earth.

The distribution of the galaxies remains the same out to a distance of 1.2×10^9 light-years — no edge to the universe is within sight. Anyhow, it is difficult to imagine an "edge" to the universe with nothing beyond. Perhaps the universe has no end of border and is infinitely large. But if this were true, then there would be an infinite number of galaxies, giving out an infinite amount of light. The brightness of a galaxy at distance $2R$ from us is one-fourth that of a galaxy at distance R. But since the galaxies are uniformly distributed, there are four times as many galaxies in a spherical shell of radius $2r$ and, let us say, one light-year thick, as there are in a similar shell of radius R. Remember, the areas of the spheres are $4\pi r^2$ and $4\pi(2r)^2 = 4(4\pi r^2)$. Because the amount of light that reaches us from one galaxy is inversely proportional to R^2 and the number of galaxies supplying the light is directly proportional to R^2, successively larger spherical shells are of equal brightness. A sphere of infinite radius would contain an infinite number of shells and they would send us an infinite amount of light. How can the sky be dark at night if we are always being supplied with an infinite amount of light? Does this mean that the universe does have a border?

Whether or not the universe is infinite, one thing we are sure of: It is not a motionless universe. Around many stars in many galaxies, rotating planets may be revolving in orbits; around many of these planets moons may circle. Each of the stars is moving around the center of its galaxy. Some of the galaxies are in pairs, moving around their centers of mass; others are members of clusters of galaxies, moving in complicated orbits around each other and around the cluster center. In addition to these motions, V. M. Slipher at Lowell Observatory in

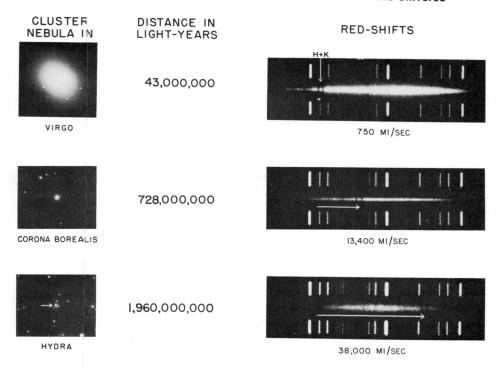

CLUSTER NEBULA IN	DISTANCE IN LIGHT-YEARS	RED-SHIFTS
VIRGO	43,000,000	750 MI/SEC
CORONA BOREALIS	728,000,000	13,400 MI/SEC
HYDRA	1,960,000,000	38,000 MI/SEC

Fig. 25-1 In the left column are photographs of successively more distant galaxies. In the right column is the spectrum of each of these, showing the Doppler shift of two strong absorption lines due to ionized calcium. The arrows indicate the displacement of each line toward the red end of the spectrum (right). Below each spectrum is the speed in miles per second that each red-shift indicates, according to Doppler's formula, $v/c = \Delta\lambda/\lambda$. The distances of the galaxies are shown in the central column. For the Virgo galaxy the distance is determined by luminosity comparisons of objects within the galaxy (p. 190); for other galaxies distance is determined by assuming that all galaxies are about equally luminous and that $b \propto 1/r^2$. [Mount Wilson and Palomar Observatories.]

Arizona found in 1912 that most of the galaxies, or "clouds" as they were then called are moving *away* from us (p. 187), some with speeds as high as 1125 miles per second. In 1929 Hubble measured the velocities of all the galaxies whose distances had been reliably estimated. He found that the galaxies and clusters of galaxies are all moving away from us (except for a few nearby ones), and that their speeds v are directly proportional to their distances R from us. That is, $v \propto R$, except for a few nearby galaxies. A galaxy 10^6 light-years from us is receding at 20 miles per second, one at 2×10^6 light-years distance is receding at 40 miles per second, and so on (Fig. 25-1). The relationship between distance and speed holds out to 8×10^7 light-years, the farthest that galaxy distances can be measured from the brightnesses of cepheids, giant stars, clusters, or novae (p. 190). It also applies to more distant galaxies where distance estimates are based on the relative brightnesses of whole galaxies—out to 2.6×10^9 light-years, as far

197

as galaxies can be photographed with long time exposures. This velocity-distance law is written $v = kR$. When v is in miles per second and R is in units of 10^6 light-years, the value of k is 20 (mi per sec) per million light-years (Fig. 25-2).

Since velocity is determined by the amount of Doppler shift of the lines in a galaxy's spectrum and since this shift is toward the red end of the spectrum, the equation $v = kR$ has become known as Hubble's "law of red-shifts." Nowadays it is used as a means of determining galaxy distances ($R = v/k$), since the speed of recession can be measured for any galaxy more simply than its brightness. If the Doppler shift in a galaxy's spectrum indicates that the galaxy is receding from us at 40,000 miles per second (by $v = \Delta\lambda/\lambda$), its distance is then $R = v/k$ $= 4 \times 10^4/20 = 2000$ million light-years $= 2 \times 10^9$ light-years.

Until 1963 the largest red-shift measured was 76,000 miles per second, for a galaxy in a very distant cluster. Then Maarten Schmidt of Palomar Observa-

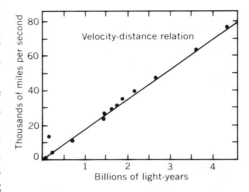

Fig. 25-2 Even in the remotest galaxies that can be observed today, there is a nearly straight-line relationship between the shifts and the distances of galaxies ($v \propto r$). So far, red-shifts have been observed for almost a thousand galaxies. Measurements of the speeds of recession indicate that the formula can be written $v = kr$, where $k = 20$ miles per second per million light-years. That is, a galaxy 10^6 (a million) light-years away is moving at $v = 20$ miles per second; one at 100 million light-years is moving at (100) (20) = 2,000 miles per second, and so on. [Diagram by Thornton Page, courtesy of *Sky and Telescope.*]

tory began to study the spectra of faint blue "stars" at points in the sky from which strong radio emission is coming. He found that these "stars" have very large red-shifts, indicating that they are moving away from us at speeds up to 100,000 miles per second. According to Hubble's law ($v = kR$), this means that they are extremely distant — up to several billion light-years away. To look as bright as they do, their luminosity must be over a hundred times that of the most luminous galaxies. Yet their angular diameters show that they are not giant galaxies, but actually smaller than most galaxies — about 25,000 light-years or so in diameter. They have emission-line spectra, which indicates that their outer layers are low-density gas. This means that their masses are probably less than a million suns and, therefore, much less than the mass of an average galaxy. Their energy output, however, is at such a rate that in 50,000 years or less all the nuclear energy available in such a small mass would be used up. In fact, there are no known nuclear reactions that can produce energy at such a high rate from so small an amount of matter.

These quasi-stellar objects (meaning "starlike" objects) are known as QSO's or quasars. Without a large telescope they cannot be seen; without radio studies they would not have been noticed especially; without the Hubble law their distances would not be known and they would be assumed to be smaller, nearer masses of gas. Are they really so far away? Their distances are determined only by the Hubble law, $R = v/k$. If this law does not hold for quasars and they are actually much closer, then they are not as luminous as we think, their energy output is not as great, and they are smaller. However, astronomers have no reason to doubt the Hubble law, and they can find no other explanation of the quasars' huge red–shifts. So it is now considered that they are very distant, and that their high luminosities are due to violent and short-lived explosions of oversized stars in galaxies of an odd type. The large, violently exploding star becomes so luminous that it alone can be seen for a few hundred light-years. As you might expect in such an explosion, the quasars are variable in brightness. Another explanation is that a quasar may be a collapsing galaxy — imploding, rather than exploding. Collapse (caused by gravitational self-attraction) is prevented in normal galaxies by rotation (p. 181) and in normal stars by gas pressure (p. 145). The problem of the quasars is still unsettled. Arguments about it wax with the vigor of the older arguments: How was the solar system formed? What makes some stars variable? Are faint oval "clouds" really distant galaxies? Observations of quasars are occupying many astronomers today. New theories about them may have been suggested, or "proved," by the time you read this book. One thing we can be sure of: The answer to the question of the quasars will raise more questions. It will also lead to differences in our picture of the universe, just as each new explanation has done in the past.

There is one thing about Hubble's discovery ($v = kR$) that may be bothering you. Does the fact that all the galaxies are moving away from us mean that our Milky Way galaxy is at the center of the universe? Cheated out of having our earth at the center of the universe, then of having our sun at the center of the Galaxy, are we at last to find ourselves with the dubious honor of belonging to a galaxy so repellent that all the others are fleeing from it, so that in time the Milky Way galaxy will be alone at the center of the universe?

The answer to this question is shown in Figure 25-3. Let us consider four galaxies, 1, 2, 3, 4, equally spaced along one line. (We must also imagine other galaxies above and below this line and above and below the plane of the paper, extending in all three dimensions.) The arrows in Figure 25-3a illustrate Hubble's law as seen from galaxy 2 (which we may imagine as the Milky Way galaxy). What does an observer on galaxy 3 see? This is shown in Figure 25-3b. He would think, of course, that his galaxy is at rest and that galaxies 2 and 4 are moving away from him on either side. He would see galaxy 1, twice as far away, moving twice as fast as galaxy 2. The same reasoning applies to an observer on galaxy 1 or galaxy 4, or on all the others not shown in the figure. Unless we can tell who is moving and who is at rest, everyone gets the same view of the others. Physicists have found no way to determine which one is at rest. The movements of galaxies do not show where the center of the universe is, or indeed, if there is a center.

Nevertheless, an observer on any of these galaxies would see that since all the others are moving away from him, at one time in the past they were all close to his galaxy. Since this is true for all galaxies, it must mean that in the past all galaxies were closer together than they are now. The universe is expanding. If we trace the motions of other galaxies back in time, we find that about 10 billion

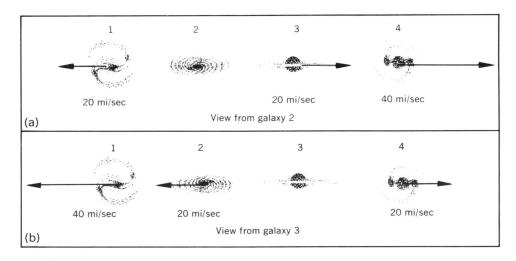

Fig. 25-3 Two views of the motions of four galaxies in a line, separated by a million light years.

years ago they were all close together (see Calculation 6). Of course, an observer on galaxy 3 (Fig. 25-3) would say that they were all near him, and an observer on galaxy 4 that they were all near him.

Thus, it looks as though all the galaxies were close together 10 billion years ago, but are moving apart like the fragments from an exploding bomb. The fragments that are moving faster—the faster galaxies—have gone farther than the slower ones. This interpretation is called the "Big-Bang" theory (Fig. 25-4a, b). The time from the initial explosion— 10 billion years—is the age of the universe. The ages of earth's oldest rocks, the ages of the oldest stars (Table 18), and the ages of the oldest star clusters (p. 175) fit this maximum age fairly well. Since the view from any galaxy is the same (Fig. 25-3), each galaxy thinks that it was the center of the explosion. So there is no answer to the question, "Where was the center?"

As time goes on, stars form from the gas clouds in all the galaxies. Eventually, when all the hydrogen has been converted to heavier elements, the supply of nuclear energy will be exhausted. The galaxies will be much more widely separated (Fig. 25-4b) and dim because all their stars are dead. The universe, according to the Big-Bang theory, will then be very different than it is today.

This sounds logical. But is it necessarily right? If there is no center to the universe (that is, if no galaxy has a preferred position and if the universe looks the same from wherever you view it), is it not possible that it looks the same whenever you view it? The "Steady-State" theory, competing with the Big-Bang theory, does not suggest a sudden beginning 10^{10} years ago. This suggests instead that the universe has always looked the same and always will. It could, if new matter is being created at the rate of about one hydrogen atom per cubic mile per year, so that new galaxies are formed to replace old ones moving away from us and

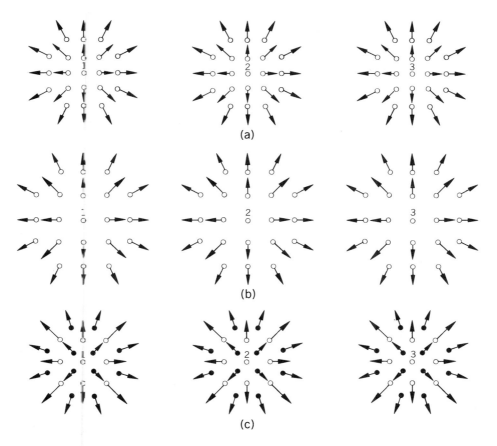

Fig. 25-4 (a) The view of the universe from galaxy 1 is identical to that from galaxies 2 and 3 — and from any other galaxy in the universe. Each dot represents a galaxy, each arrow a red-shift velocity, larger at greater distances. Millions of receding galaxies should, of course, be pictured over an area hundreds of times the size of this page, and also above and below the plane of the paper. (b) Five billion years after the time shown in (a), according to the "Big-Bang" theory, the view from galaxies 1, 2, and 3 is like this. The retreating galaxies are more widely spread out. The view from all galaxies is identical, but it is different from that shown in (a) at the earlier time. (c) According to the "Steady-State" theory, however, the view from galaxies 1, 2, and 3 will not only look the same for all after five billion years, but it will look just about like it did at the time shown in (a). This is because twelve new galaxies (shown by solid circles) have formed during the five billion years.

from all the other galaxies (Fig. 25-4c). The universe would then remain at the same density, and there would always be young galaxies. In 10 billion years it would not be a starless universe as the Big-Bang theory predicts.

How can these rival theories be tested? According to the Big-Bang theory, the most distant galaxies should be younger ones. They are farther away, so their light takes longer to reach us, and we see them as they looked long ago when they were younger. We see a galaxy 10^6 light-years away as it looked 10^6 years ago; we see one 10^9 light-years away as it looked 10^9 years ago. If they all formed at once, then we are seeing a snapshot of one when it was only 10^6 years younger than the Milky Way; and of the other when it was 10^9 years younger, only nine billion years of age, rather than 10 billion, as we are. According to the Steady-State theory, the average age of galaxies is about the same whatever their distance: Some are young, some are middle-aged, and some are old. But in order to use these two predictions to test the theories, astronomers must know how to tell a young galaxy from an old one. And as yet they do not.

Today, astronomers are arguing the merits of the two theories as vigorously as once the Ptolemaic and Copernican theories were argued. They are searching for evidence, as Tycho and Galileo searched. In our lifetimes they may find it. Then will astronomy be finished? From past experience it seems sure that the astronomers will go back to their telescopes to find answers to questions we have not thought of yet. They will find that the universe is even more complex than we now think—just as Ptolemy, as he drew his epicycles and watched the celestial sphere rotate, never dreamed that the earth, center of his universe, would one day be found to be merely one of nine planets circling one of the billions of stars in one of the billions of galaxies in the universe.

Never would he have dreamed of today's huge radio telescopes measuring invisible radiation from stars and nebulae, or listening for the sound of other voices broadcasting from the planets of other stars—stars in our galaxy, or stars in any of the others, which are spread out in every direction as far as the instruments that man has made can see or hear.

We, on the earth, are like passengers on a small spaceship in heavy traffic, revolving around our sun, which carries us with it in a slow path around the center of our galaxy, which in turn is moving through the universe, who knows where, in the company of billions like it. The jewels that light the night sky are not ours alone; we very likely share them with many another observer on many another planet. These planets had—or will have—their Ptolemys and Galileos and Newtons, their Herschels and Hubbles. They may also have students like you, who have patiently worked your way through the calculations, figures, tables, and text of a book like this, in order to learn a bit about your surroundings. Somewhere, out on some distant galaxy, students may be having as much trouble with the ideas that we call Doppler shift, or center of mass, or Kepler's law.

Now when you look at the night sky, besides appreciating its sparkling beauty, perhaps you feel pleased at how much men have been able to learn about the universe by patiently observing the silent sky and thinking clearly about what they saw. Perhaps you have shared some of the excitement astronomers felt as they solved so many of its mysteries; perhaps you are curious to see what the answers will be to the many unsolved problems which remain—or perhaps you will help to find the answers to them.

Additional Reading

ABELL, G. O., "Galaxies: Landmarks in the Universe" in *Stars and Galaxies* (Thornton Page, ed.): Englewood Cliffs, N. J., Prentice-Hall, 1962.

HOYLE, FRED, "When Time Began" in *Astronomy* (Samuel Rapport and Helen Wright, eds.): New York, New York University Press, 1964.

HUBBLE, EDWIN, *The Realm of the Nebulae:* New York, Dover Publications, Inc., 1958.

Appendix

Calculation 1

In astronomy some very large and some very small numbers are used. It is tedious to write out all the zeros, or to say "nine hundred billion" or "one-nine-hundred-billionth." Therefore, the following system called *powers of ten,* is used.

(1) The gravitational constant G is 0.0000000668. This is written 6.68×10^{-8}.

$$0.001 = 1/1000 = 1/10^3 = 10^{-3}$$
$$0.006 = 6/1000 = 6/10^3 = 6 \times 10^{-3}$$
$$0.0066 = 66/10,000 = 66/10^4 = 66 \times 10^{-4} = 6.6 \times 10^{-3}$$
$$0.0000000668 = 6.68/10^8 = 6.68 \times 10^{-8}$$

(2) The mass of the earth is 6,600,000,000,000,000,000,000 tons. This is written 6.6×10^{21} tons.

$$1000 = 10^3$$
$$6000 = 6 \times 10^3$$
$$6600 = 6.6 \times 10^3$$
$$\text{Mass of earth} = 6.6 \times 10^{21}$$

(3) To multiply, multiply the figures and add the powers of ten:

$$(6 \times 10^4) \times (2 \times 10^2) = 12 \times 10^6 \text{ or } 1.2 \times 10^7$$

(4) To divide, divide the figures and subtract the powers of ten:

$$(6 \times 10^4) \div (2 \times 10^2) = 3 \times 10^2$$

(5) To add or subtract:

$$(6 \times 10^4) + (2 \times 10^2) = (600 \times 10^2) + (2 \times 10^2) = 602 \times 10^2 = 6.02 \times 10^4.$$
$$(6 \times 10^4) - (2 \times 10^2) = (600 \times 10^2) - (2 \times 10^2) = 598 \times 10^2 = 5.98 \times 10^4.$$

Calculation 2

F = force of gravity between the earth and the apple.

m = mass of a falling apple (or any other object).

M = mass of the earth.

a = acceleration of the apple (or of any other freely falling object) toward the earth.

R = distance of the apple from the earth's center (= radius of the earth).

$$F = ma = \frac{GmM}{R^2}$$

The two m's cancel out:

$$a = \frac{GM}{R^2}, \; GM = aR^2, \text{ and therefore, } M = \frac{aR^2}{G}.$$

Since we know a (32 ft/sec, each second) and r (4000 miles), as well as the value of G, we can solve for M, the mass of the earth.

Calculation 3

Problem: To find the mass of the sun (M).
We can measure or calculate:
  a = acceleration of the earth in its orbit (cm/sec²).
  v = earth's velocity in its orbit around the sun (cm/sec).
  R = distance of the earth from the sun-radius of earth's orbit (cm).
  P = earth's period, the number of seconds it takes the earth to go around the sun.
  G = gravitational constant (dyne cm²/gm²).
We will consider the earth-moon system, of mass m, to be going around the sun in a circular orbit. Formulae for an elliptical orbit are more complicated.
Acceleration a in a circular orbit is given by the following formula:

$$a = \frac{v^2}{R}$$

The velocity v around the orbit is the circumference $2\pi R$ divided by the period P:

$$v = \frac{2\pi R}{P}$$

If we substitute v from this formula in the first formula, we get:

$$a = \frac{(2\pi R)^2}{P^2} \times \frac{1}{R} = \frac{4\pi^2 R^2}{P^2} \times \frac{1}{R} = \frac{4\pi^2 R}{P^2}$$

Now F, the force of gravitation between the earth and the sun, is accelerating the earth's mass. Therefore:

$$F = ma$$

Substituting a:

$$F = (m)\left(\frac{4\pi^2 R}{p^2}\right) \quad \text{or} \quad F = \frac{m4\pi^2 R}{P^2}$$

F is also the force of gravitation between the earth and the sun. Therefore:

$$F = \frac{GmM}{R^2} \quad \text{and}$$

$$\frac{m4\pi^2 R}{P^2} = \frac{GmM}{R^2}$$

Canceling m and cross-multiplying gives M (mass of the sun) in terms of the measured quantities R, P, and G:

$$(m4\pi^2 R)(R^2) = (P^2)(GmM)$$

$$4\pi^2 R^3 = P^2 GM$$

$$M = \frac{4\pi^2 R^3}{P^2 G}$$

Calculation 4

E = energy emitted by the sun (in calories per sq cm per min).
T = temperature in degrees K.
k = a constant, 76.8×10^{-12}.

$$\text{Stefan's law: } kT^4 = E$$
$$(76.8 \times 10^{-12}) T^4 = 89{,}676$$
$$T^4 = 89{,}676/76.8 \times 10^{-12}$$
$$T^4 = 1168 \times 10^{12}$$
$$T = 5845°K$$

Calculation 5

$$\lambda_{max} = \frac{k}{T}$$

$$4750 = \frac{2.897 \times 10^7}{T}$$

$$T = \frac{2.897 \times 10^7}{4750}$$

$$T = .00061 \times 10^7$$
$$T = 6100°K$$

Calculation 6

The Hubble constant is usually expressed as 20 mi/sec per million light-years.
  1 light-year = 6×10^{12} miles
  1 year = 3×10^7 seconds
To have once been close to our galaxy, a galaxy now 10^6 light-years away must have moved 10^6 light-years. How long did it take to move this distance?
  t = time
  R = distance
  v = speed

$$t = R/v$$
$$t = (10^6)(6 \times 10^{12}) \text{ miles}/v$$
$$v = 20 \text{ mi/sec}$$
$$t = (10^6)(6 \times 10^{12})/20 = (10^6)(3 \times 10^{11}) = 3 \times 10^{17} \text{ sec}$$
$$t = 3 \times 10^{17}/3 \times 10^7 = 10^{10} \text{ years (1 billion)}$$

If the galaxy is twice as far, it moves twice as fast. Then:
$$t = (2)(3 \times 10^{17})/(2)(3 \times 10^7) = 10^{10} \text{ years.}$$
If it is 100 times as far away, it moves 100 times as fast. So:
$$t = (100)(3 \times 10^{17})/(100)(3 \times 10^7) = 10^{10} \text{ years.}$$

Index

aberration of starlight, 62–64

absolute zero, 97

absorption (dark spectral) lines, 116–17, 120–21, 140–42, 180, 183

acceleration of gravity, 57–60, 61

Adams, J. C., 80–82

Algol, 156–57, 164

Almagest, 10, 14, 166

Alpha Centauri, 64–65, 148–49, 155

Andromeda galaxy (M31), 189, 190, 191, 195

Apollo 11 and 12 moon landings, 136

Aristarchus, 26, 38

Aristotle, 34, 36, 38, 53–54, 123, 164–65

asteroids (minor planets), 124–27

astronomical unit (AU), 44, 73

Barnard's star, 176, 183–84

Bessel, F. W., 64, 148

Big-Bang theory, 200–202

Big Dipper, 7–8, 12, 148, 155, 156, 183

Bode, Johann, 77, 79, 81, 124

Bode's law, 124, 152

Bradley, James, 62–64

Brahe, Tycho, 34–43, 61, 64, 79, 94, 148, 164

Cannon, Annie J., 151, 180

Cavendish, Henry, 61, 73, 111

celestial equator, 16, 28; poles, 13, 16, 27–28

celestial sphere, 9–11, 16, 27, 39, 65, 66, 182

cepheids, 159–64, 166–67, 181, 190

clusters of stars, 51, 152ff, 156, 174ff, 181, 186

constellations, 7–11, 18–19, 39

comets, 36–37, 38, 129–31

Copernicus, Nicholas, 26–34, 61, 94, 148

Curtis, H. D., 189

diffraction, 92; grating, 93

Doppler shift, 118–22, 128, 140, 144, 159–60, 183–85, 187, 190, 195, 197–98

Draper, Henry, 83

dust, interstellar, 179–80, 194; moon, 135–36

earth, distance from sun, 73; rotation, 27, 65–68; mass, 73, 74, 204; seasons, 16–17, 28–29; shape, 11, 23; *see also* Planets

earth–moon system, 71–72, 132

eccentricity, 44–45

eclipse, lunar, 22–23; solar, 21–23, 142–43

ecliptic, 15–16, 28

Eddington, Sir Arthur, 160–62, 163

Einstein's energy–mass relation ($E = mc^2$), 146–47

ellipse, 43–44

emission (bright spectral) lines, 128, 130, 141–44, 187, 198

epicycles, 25–26, 32, 39, 42, 45

escape velocity, 111–15

Foucault pendulum, 65–67

Fraunhofer, Joseph, 115–16, 120, 151

frequency, 121

galaxies, 186–202; changes in, 194; clusters, 194, 196; diameters, 190, 191, 196; distances, 189–90, 191, 196; distribution, 187, 195–96; evolution, 191, 194; formation, 201; masses, 190, 191, 196; number, 191, 195; orbits, 196; pairs, 190, 196; radial velocities, 187; recession, 187, 197ff; rotation, 190, 194; spectra, 187, 197; types, 191ff

galaxy, Milky Way, 178–85, 199; arms, 185; center, 181; diameter, 181–82, 190; mass, 185, 190; motion, 195; nucleus, 181, 188; rotation, 184–85; stars, 185; shape, 177, 182; type, 194

Galilei, Galileo, 48–57, 61, 66, 74, 122, 123, 132, 136, 177, 186, 202

Goodricke, John, 156–57, 159, 164

gravitational constant (G), 73

gravity, 57–61, 68, 110–11

Henderson, Thomas, 64, 148

Herschel, Caroline, 150, 155

Herschel, Sir John, 186

Herschel, Sir William, 75–79, 94–95, 98, 150, 155–56, 159, 178–79, 182, 186, 202

Hertzsprung–Russell (H–R) diagram, 153–54, 164, 168, 176

Hubble, Edwin, 189–90, 191, 195–97

Huygens, Christian, 87, 91, 122

inertia, principle of, 56–57, 58, 66, 161

interference, 91

interstellar material, 179–80, 194

inverse-square law, 60–61

Jupiter, clouds, 106; mass, 74; moons, 53, 64, 74; motion, 32; orbit, 32; path in sky, 33; red spot, 105–106; temperature, 107, 109; *see also* Planets

Kamp, Peter van de, 176, 184

Kepler, Johannes, 41–48, 53, 57, 60, 61, 75, 94

Kepler's laws, 45–47, 57, 60, 157, 182–85, 190

Kirchhoff, Gustav, 116, 128, 187

Leavitt, Henrietta, 162

Leverrier, U. J. J., 81–82

light, nature of, 85–93; velocity, 64, 96, 149

light-year, 149, 185

Lowell, Percival, 82–83, 117–18

luminosity, 149, 150–51, 153–54, 157, 169

Magellanic Clouds, 162–63, 191, 194–95

magnitude, 149

main-sequence stars, 153–59, 167–69, 175

Mars, albedo, 104; apparent motion, 30–31, 40, 71–72; atmosphere, 108–109, 135; canals, 118; craters, 105, 109, 135; distance from earth, 73, 108; distance from sun, 32, 44; equator, 109; life, 117–18; mass, 75; moons, 75; motion, 29, 121; orbit, 31–32, 44–45; polar caps, 108–109; rotation, 121;